AF825882

Contents

India's Energy Transition

Published by
ACADEMIC FOUNDATION
in association with

Institute of Public Enterprise

India's Energy Transition

Possibilities and Prospects

Ram Kumar Mishra

V. Balaji Nagendra Kumar

ACADEMIC FOUNDATION

NEW DELHI

www.academicfoundation.com

First published in 2012
by

ACADEMIC FOUNDATION
4772-73 / 23 Bharat Ram Road, (23 Ansari Road),
Darya Ganj, New Delhi - 110 002 (India).
Phones : 23245001 / 02 / 03 / 04.
Fax : +91-11-23245005.
E-mail : books@academicfoundation.com
www.academicfoundation.com

in association with

Institute of Public Enterprise, Hyderabad

© 2012. Academic Foundation.

ALL RIGHTS RESERVED.
No part of this book shall be reproduced, stored in a retrieval system,
Or transmitted by any means, electronic, mechanical, photocopying,
recording, or otherwise, without the prior written permission of the
copyright holder(s) and/or the publishers.

Cataloging in Publication Data--DK
Courtesy: D.K. Agencies (P) Ltd. <docinfo@dkagencies.com>

Mishra, R. K., 1948-
India's energy transition : possibilities and prospects / Ram
Kumar Mishra, V. Balaji Nagendra Kumar.
p. cm.
Includes bibliographical references (p.).
ISBN 13: 9788171889730
ISBN 10: 8171889735

1. Energy development--India. 2. Energy policy--India. 3.
Energy industries--India. I. Nagendra Kumar, V. Balaji, joint
author. II. Title.

DDC 333.790954 23

Typeset by Italics India, New Delhi.
Printed and bound in India.

List of Tables and Figures

TABLES

FIGURES

About the Authors

Ram Kumar Mishra is Director, Institute of Public Enterprise (IPE), Hyderabad. Prof Mishra has numerous achievements to his credit. He is the member of the Task Forces to improve the performance and management in public enterprises affiliated to numerous Ministries of the Government of India. In the past, he was member of several UN committees, and committees formed by the Government of India towards strengthening the public sector. He is also a great teacher and a prolific writer. He has taught in various national and international universities, including the London Business School, the Sciences Po (Paris), SDA Bocconi (Milan), etc. He has contributed research papers to international and national journals, and chapters in edited volumes numbering several hundreds. Prof Mishra has guided over 60 PhD students in the field of economics and management.

V. Balaji Nagendra Kumar is the National Doctoral Fellow at the Institute of Public Enterprise (IPE), Hyderabad. He has contributed research papers in several edited volumes and journals. He was honoured with the much coveted Lee Kuan Yew Fellowship by the Lee Kuan Yew School of Public Policy (National University of Singapore) for the completion of his Masters in Public Policy. He also has M.Phil degree from the Jawaharlal Nehru University (Delhi) and MA from the Gokhale Institute of Politics & Economics (Pune). Currently, he is pursuing his doctorate at IPE on the topic 'Determinants of Solar Energy Scenario in India's Quest for Energy Security'.

Acknowledgements

We are indebted to several of our colleagues, external libraries and our families for enabling us to successfully complete this book.

Our special thanks go to Ms Y.V. Sujana, the Programmer at the Institute of Public Enterprise (IPE) Hyderabad and Ms Priyadarshini Gandhi for the technical assistance.

We are indebted to our Librarian, Ms Sunita Murthy and other library staff, including the book bearer Mr Badruddin for patiently helping us in searching for, and obtaining the books. We are also indebted to the library staff of the University of Hyderabad and the Centre for Economic and Social Studies (CESS), Hyderabad.

We specially thank Librarians Mr Varadarajan and Mr R.P. Rao of the University of Hyderabad, and Mr Sudhakar of the CESS, Hyderabad.

We thank the Ministries of Power and New and Renewable Energy, Government of India, for sharing their Strategy Blueprints, Annual Reports and for interacting with us.

We would also like to thank the Editorial Department of Academic Foundation and especally Mr Sanu Kapila for extending all possible support in bringing out this volume.

We also specially thank the IPE for providing the financial grant for the field work.

Finally, we thank our families for encouraging and supporting us while working at home towards the completion of the book.

Preface

India's energy problems are complex, multidimensional, and difficult to solve. India's transportation sector, as well as the power sector face tremendous challenges to sustain themselves. India's power shortage is about 15 per cent compared to the contemporaneous demand, and about 70 per cent when compared to the electricity use in the middle income countries *viz.*, over 3,000 KW per annum. Hence, 70 per cent of India's transportation fuel has to be imported.

The above scenario is getting translated into higher prices, and greater shortages in power and in transportation fuel *viz.*, petrol, diesel, natural gas etc. Unless long-term corrective steps are taken, India is in danger of importing almost 100 per cent of transportation fuel by 2030, and is even in danger of using coal for transportation purpose. If coal is used for transportation, its availability for power generation will shrink faster, and it will also cause enormous amount of pollution. Incidentally, 2/3rd of power generation in India is through the thermal source. This is likely to remain so in the foreseeable future.

Government of India primarily focusses on the power sector to take corrective measures. Corrective measures include: energy conservation, improving energy efficiency in buildings, promotion of the use of renewable energy, including the nuclear, solar and wind power, and improving the plant load factor (PLF) in the power generation industry.

Overtime with the depletion of fossil fuels *viz.*, oil, natural gas and coal, renewable energy assumes important status in power generation.

The recent trends suggest that electricity is becoming indispensable for transportation as well. The recent introduction of fuel cell/hybrid vehicles, the future promises stored in the upcoming solar energy technologies, and the advances in the battery storage technology is to make electricity a perfect substitute for fossil fuels in transportation.

This book tries to present the possibilities and prospects for India's energy transition. Chapter 1 introduces the energy economics, energy security, the energy resources, the energy reserves, and discusses the energy use patterns in the developed and developing countries. Chapter 2 discusses the energy security in the international, national and regional contexts in detail. The chapter also discusses various energy choices and options, including the renewable, nuclear etc. Chapter 3 presents the rural-urban energy consumption pattern, the coexistences of traditional forms of energy with modern forms led by fossil fuels and electricity, presents trends in oil and natural gas production and imports, and the trends in electricity generation, use and deficit. The Chapter also discusses the efforts made by the Government of India to bolster domestic reserves, and production of oil and natural gas through various rounds of New Exploration Licensing Policy (NELP). Chapter 4 discusses the energy efficiency measures, and contextualises in the climate change scenario.

Chapter 5 discusses the trading in oil and natural gas, various pricing formulas, various markets, types of fuels traded etc. Chapter 6 discusses the power trading, the trading instruments *viz.*, derivatives like the forwards, the futures, and options and their forms, and even the financial derivatives used in power exchanges. India's recent drive towards power trading is discussed along with various electricity markets. Chapter 7 justifies India's quest for nuclear energy. Chapter 8 discusses the relevance of renewable energy in mitigating the adverse effects of climate change, the development and patenting patterns, and international distribution of forms of renewable energy, patenting focus etc. Chapter 9 discusses India's performance in the renewable energy sector *viz.*, wind, solar, small hydro, biomass etc. Wind power dominates India's renewable energy scenario which could be overtaken by the solar power after the year 2020. Chapter 10 presents the solar energy trends, viability, international technology and import-export dimensions, India's position, the Jawaharlal Nehru National Solar Mission and the accompanying power, skill, technology and strategy development. Besides, solar subsidy schemes of some states of India are also discussed. Chapter 11 discusses the future thrust of India's power sector, ways of strengthening the generation, transmission and distribution sectors, enhancement of equipment manufacturing and technology development in India. Besides, the chapter identifies the pitfalls in the power project construction, and discusses the means of overcoming them using the project management method. Chapter 12 presents the concluding remarks.

1 Introduction

Human society cannot survive without a continuous stream of the supply of energy resources and energy commodities.

In physical science terminology, 'energy' is the capacity for doing work, e.g., lifting, accelerating, or heating material. In economic terminology, 'energy' includes all energy commodities and energy resources, commodities or resources that embody significant amounts of physical energy and thus offer the ability to perform work.

Energy commodities *viz.*, petrol, diesel, natural gas, propane, coal, electricity etc., can be used to provide energy services for human activities, which include lighting, space heating, water heating, cooking, motive power and electronic activity. Energy resources *viz.*, crude oil, natural gas, coal, biomass, hydro, uranium, wind, sunlight, geothermal deposits etc., can be harvested to produce energy commodities.

This chapter introduces the concept and classification of energy resources, the present mode of energy consumption *viz.*, the fossil fuel and bioenergy-based consumption, lack of universal access to fossil fuels, rapid depletion of fossil fuel and firewood and dung and then the need for a more sustainable mode of energy consumption *viz.*, the renewable energy.

The need for prolonging the availability of fossil fuels through appropriate pricing and utilisation policies and regulatory mechanism is recognised. Then, the need for developing, promoting and using the energy conservation measures is recognised and emphasised.

Classification of Energy Resources

Depletable *versus* Renewable Resources

Based on the speed of natural process one can classify primary energy resources as depletable or renewable. Depletable resources *viz.*, the coal, oil and natural gas take millions of years to form again in the earth's core.

Whereas, renewable resources get replenished much quickly. However presently, the cost of harnessing depletable resources is less than the cost of harnessing renewable resources. Renewable resources can be further subdivided into storable or non-storable resources. Storable resources are derived from a stock, which can be used or stored. Biomass, hydropower and some geothermal come under this category. Storable form of renewable energy is depletable in short term and renewable in medium to long term. Non-storable resources *viz.*, solar, wind and hydro can be used or not used without influencing the harnessing and usage in future time period.

At present, the dominant source of energy is depletable resources and is likely to be so for the next 20 years or more. While the dependence on renewable energy is inevitable, the questions of which type of renewable energy would dominate, timing of a complete transition to renewable energy and the ways and means of efficiently achieving such a transition need to be fully ascertained.

The Background

History reflects human being's endeavour to control over energy sources for the benefit of society. Energy consumption has become an important indicator of a country's modernisation. Conventional energy resources *viz.*, oil and natural gas are largely concentrated in few countries and regions in the ocean necessitating large-scale imports by many rich and developing countries.

Cost of oil and natural gas and issues of geopolitical security have become important concerns. This apart, increasing environmental hazards and the need to control these hazards necessitate the control of the use of conventional fuels, improvement in energy conservation, the development of eco-friendly technologies and the harnessing of energy from renewable sources *viz.*, sunlight, wind and water etc.

World's oil and natural gas stocks would last only for about 40-60 years and coal for about 150-200 years respectively. Besides, the oil and natural gas cartels *viz.*, the Organization of Petroleum Exporting Countries (OPEC) etc., controls the prices whose unforeseen fluctuations cause financial and material losses to rich as well as to the poor countries. The sudden three-fold and two-fold increases in oil price during the 1973 and

1979 oil shocks and the recent rise of oil price to about $150 reflect a grim scenario.

The gloomy prognostication of oil and natural gas is in spite of the fact that about a third of world's population *viz.*, two billion people are depending on traditional sources of fuel *viz.*, firewood, cow dung etc., whose supply itself is precarious. Social forestry and livestock population are dwindling alarmingly. The increased rate of urbanisation will only add to the gloomy prognostication concerning oil and natural gas. Energy concerns extend beyond the questions of availability and cost. Acid rains, buildup of carbon dioxide in the atmosphere and the accompanying increase in global warming, nuclear weapons proliferation and geopolitical security of energy transportation are partially or wholly rooted in energy.

The Influence of Oil on the Economic Base of Industrialised Countries

Between 1925 and 1972 (the year before the great oil shock), energy budgets of industrialised countries witnessed a great shift. The countries depended on coal by over 80 per cent in 1925. By 1972, the industrialised countries depended by up to 70 per cent on oil and natural gas. Oil consumption grew from 2 to 45 million barrels per day at the spectacular average annual rate of 6.5 per cent per year. Oil is ubiquitous. Hence, its remarkable properties are taken for granted. Oil has a high energy density, is easy to transport, store and use, and it is much cleaner than coal, for many decades, oil was incredibly cheap. In 1970, five US gallons (20 litres) of oil cost only $1.0 (in 1984 price) at the well-head. Nature's manufacture of oil made the marginal cost of oil quite low.

Over time, the cost of discovery, extraction, processing and distribution of oil came down further. Consequently, oil prices remained stable and even declined until 1973. The modern lifestyle of comforts and luxury got manifested in the form of easy private mobility of people, faster and bulk movement of freight, consumption of foreign and out of season fresh fruits and vegetables, economical central heating and air cooling facilities etc., due to the proliferation of the use of oil. Only the 1973 and 1979 oil shocks resulted in people's quest for sustainable lifestyle and betterment of living standards using sustainable forms of energy.

The Influence of Oil on the Economic Base of Developing Countries

In developing countries, industrialisation in its true sense began after World War II. Until then, traditional biofuels *viz.*, the firewood, charcoal and animal residue was the primary source of fuel. Increased use of automobiles for passenger and freight transport and increased use of kerosene and LPG in cooking shifted the pattern: especially in Brazil and in India. In Brazil for example, petroleum accounted only for about 9 per cent of total energy consumption in 1940 which increased to about 50 per cent by 1975.

The Impact of the Oil Shocks

In late 1973, Arab oil producers, reacting to US' steps to re-supply Israel during the October 1973 Arab-Israeli War, embargoed oil shipments to the United States and reduced oil output and shipments to other countries. For the first time, the oil-exporting countries set the price of oil obtained from their own territories without consulting the foreign oil companies involved in exploration, development, production, transportation, refining and distribution.

Oil price was increased three-fold to about $7 in 1973. In 1972, the Shah of Iran described the situation thus: 'in 1947, the posted price for a barrel of oil in the Persian Gulf was $2.17. Then it was brought down to $1.79, and that lasted until 1969. So there were 22 years of cheap fuel that made Europe what it is, that made Japan what it is. Then the price of wheat jumped 300 per cent, vegetables the same, and sugar in the past six years increased 16 times. So we charged experts to study what prices we should put on oil. Do you know that from oil you have today 70,000 derivatives? When we empty our wells, then you will be denied what 1 call this noble product. It will take you $8 to extract your shale or tar sands. So I said let us start with the bottom price of $7; that is the government intake.' A combination of three factors gave the Persian Gulf OPEC nations the leverage in 1973 that enabled the massive price hike.

(1) The increased consumption of oil, particularly by the industrialised countries,

(2) The levelling-off of US oil production, and

(3) The resulting increased shift in dependency on the oil produced in the Persian Gulf.

In 1950s and 1960s, the oil imports of the industrialised countries with market economies were far below the 1973 level and the US had sufficient spare capacity to supply its allies in the event of supply disruptions. However, oil imports by the major importers (the US, Japan and Western Europe) grew dramatically from 6 to 27 million barrels a day between 1960 and 1973. Between 1970 and 1973, US alone doubled its daily oil imports from 3 to 6 million barrels. Oil exporters from the Persian Gulf region quadrupled their oil production between 1960 and 1973 and the share of Persian Gulf oil in the industrialised market economies grew from 28 per cent to 44 per cent during the corresponding period.

From $7, the oil price grew to $22 between 1975 and 1979 due to economic recovery from 1975. The increased demand for oil served as a catalyst to OPEC to administer the second oil shock in 1979 leading to doubling of oil price to $45 between 1981 and 1984. Organisation for Economic Co-operation and Development (OECD) countries ended up spending 1/6th of their GDP on oil contributing to stark reduction in their purchasing power of other goods. On the one hand, oil price increased caused inflation, but on the other, there was a stark reduction in the demand for other goods and services resulting in the phenomenon of stagflation.

Developing countries suffered from a greater pinch in resources. Substantial amount of foreign exchange was devoted to oil imports drying up investments for development. Similarly, the export earnings of developing countries also reduced due to the reduction in the purchasing power of the industrialised countries. The Terms of Trade for developing countries also reduced due to the increased price of imported manufactured goods *vis-à-vis* the reduced price of raw materials. Increased external debt to finance the projects coupled with worsening Terms of Trade led to a large increase in the debt to GDP ratio from about 15 per cent to 34 per cent and debt servicing to export earnings ratio from about 12 per cent to 20 per cent respectively. Over 2/3rd of oil reserves are found in OPEC countries and over half of it in countries in and around the Persian Gulf.

Fuel Wood Crisis: The Ultimate Doom

Fuel wood is a major energy source in the traditional sectors of developing countries. Fuel wood use in the domestic sector (cooking, water heating and space heating, but mainly cooking) is done with traditional technologies of very low efficiency (of the order of 10 per cent) and accounts for about half the trees which are felled in the world. In addition, small-scale and rural industries such as brick-making also depend on fuel wood.

This dependence has put increasing pressure on the biomass resource base that sustains fuel wood demand. The human population which relies on this fuel has been growing exponentially. Between 1961 and 1975, fuel wood consumption (including the wood used to produce charcoal made from this fuel wood) grew at a rate of about 2.1 per cent per year. Increasingly in the developing world, fuel wood is being consumed faster than it can be regenerated. The present depleting mode of fuel wood use is not sustainable.

The human consequences of this situation are profound. The fuel wood crisis involves human suffering on a grand scale. It involves back-breaking drudgery and long hours. As the resource base dwindles, fuel wood gatherers must spend more and more time and trek ever increasing distances finding what they need. Women and children in particular must toil daily to keep their families supplied with fuel woods for cooking and heat. It interferes with the children's schooling and contributes mightily to the misery of poor women. And, amplifying this human suffering is the tremendous ecological damage arising from the forestation caused by fuel wood demand.

The Electricity Crisis

Electricity problem stems both from increasing growth in demand and supply constraints. Supply constraint is a very important issue due to high marginal cost of production using conventional sources of energy and due to infrastructural constraints in transmission and distribution of adequate quality power at all times of the day through appropriate planning and management mechanism.

India's Energy Security Concerns

India and the world as a whole need a sustainable future. Energy security, environmental sustainability and economic competitiveness are the backbone of sustainable future. Sustainable energy security holds the key to economic sustenance and economic competitiveness. Renewable energy is essential to ensure sustainable energy security because the conventional resources of energy *viz.*, coal, oil and natural gas are fast depleting.

India imports over 70 per cent of its oil requirements and over half of its natural gas requirements. In so far as the oil is concerned, India's share of imports would increase to about 90 per cent by the year 2031-32. Moreover as noted above, oil and natural gas as a whole would be exhausted in about 40-60 years and the use of coal is being discouraged even in power generation due to environmental reasons.

A complete transition to the use of renewable energy should be achieved by the time the depletable energy resources get exhausted. By the time the depletable resources get exhausted, renewable energy resources should be sustainably available, price competitive, freely tradable, legally accessible and technically feasible. The study of 'energy future' henceforth requires a proper understanding of the economic implications of energy harnessing and use, energy conservation and the legal and institutional support provided to ensure energy security, environmental sustainability and economic competitiveness.

Depletable Resource Economics

Depletable resource economics studies the economically optimal way of extracting a perfectly known stock of the resource, with perfectly predictable future prices of the extracted commodity. It essentially means that under optimal extraction path, the resource owner explicitly or implicitly recognises an opportunity cost, or rent, in addition to the marginal extraction cost.

All the information concerning the role of future prices and costs would be embodied in the opportunity cost. The perfectly competitive firm would extract the resource until the marginal extraction cost plus opportunity cost equals its selling price. Beyond this point over time, the extraction rate of depletable resources would fall drastically. As a result, the market prices would increase gradually up to the cost of producing substitutes *viz.*, renewable resources. Renewable would be produced in increasing quantity as the market price of depletables increase up to the point of complete depletion.

The timing of the transition to renewable energy is not certainly predictable due to the possible future discovery of depletables, technological advances to make the hitherto uneconomical extraction of depletables economical and due to other market information acting as a deterrent to the timely transition to renewable. Perhaps, greater amount of renewable energy itself would be available enabling the transition before the depletion of the depletable resources. Perhaps, the depletables would deplete faster than the increase in the availability of renewable energy causing scarcity of energy supply.

Before understanding the ways and means of transition to renewable energy, it is necessary to recognise the fact that depletable resources: more importantly natural gas will have their sway for the next 40-60 years. India, by and large, imports oil and natural gas. Like other major oil and natural gas importing countries like China, the US etc., India is facing sudden price fluctuations due to the increase in demand, inadequate supply infrastructure and unprotected supply roots.

The tendency of oil and gas producers to cartelise is also keeping the price high.

Strong economies like the US and China having the large consumer base and military power as their advantages ensure better terms for themselves by virtue of long-term contracts and ownership rights over exploration and production facilities.

India has less of such advantages. Yet, India can limit the number of buyers by letting large players like the Oil India Ltd. (OIL) etc., enter into contracts and agreements through appropriate regulatory mechanism. Regulatory mechanism should ensure adequate investment in hydrocarbon sector and incentives, rents and risks should be distributed throughout the value chain.

Government has to play an active role in ensuring the development and growth in the use of renewable energy. Towards this effect, Government has to play an important part in providing finance, creating conducive environment for private investments and also by helping in the realisation of the technical, economic and environmental goals by providing scientific and technological support to researchers and investors.

As a first step, pricing strategies pertaining to and taxation on depletable energy resources help provide finance for R&D and increased production and use of renewable energy.

As a second step, promotion of energy conservation measures and technologies coupled with the regulation of production and imports of oil and natural gas enhances the timing of their availability and would give adequate time for the fruition of the R&D efforts pertaining to renewable energy.

As a third step, progressive increase in the share of renewable energy in total energy consumption makes the price of energy commodities less fluctuating and more conducive for the consumers.

The Scheme

This book has 12 chapters including the introductory and concluding chapters.

Chapter 2 discusses the concept of energy security and India's relative position and role in enhancing its energy security.

Chapter 3 discusses the recent trends in India's energy scenario pertaining to the use of energy commodities by the household, India's oil and natural gas policies and finds, export and import of petroleum products, energy demand forecast scenario and the electricity deficit and losses.

Chapter 4 discusses the energy efficiency measures and the promotion of sustainable development using the National Action Plan on Climate Change. The Missions for sustainable habitat, energy efficiency and waste to energy assume added significance.

Chapter 5 discusses the energy trading with particular reference to the oil and natural gas markets. Chapter 6 discusses power trading. The need for power trading, the mechanism of power trading, the key drivers of power trading and various contracts of power trading that are practiced 'over the counter' and exchange based are discussed. Finally, power trading markets in important countries such as the United Kingdom, the Nordic Pool, the United States/California and India are discussed.

Chapter 6 discusses the various facets of power trading, including the mechanism, trading and contractual instruments, trading practices in leading power exchanges and in India.

Chapter 7 discusses India's nuclear policy, opportunities and challenges in the context of the international deals—including the Indo-US nuclear deal.

Chapter 8 discusses India's efforts towards transitioning into a renewable energy regime. The adverse impact of climate change, the need to immediately mitigate the greenhouse gas (GHG) emissions and the need to ensure sustainable development warrant the transition to renewable energy. The chapter recognises that 'technology' is the key to effect this transition. Hence, international technological efforts in terms of R&D efforts and the patent applications and achievements are discussed. Japan, the USA, Germany, China, South Korea etc., are the most important patent holders and the adapters of renewable energy. India is one of the leading adaptor of renewable energy: especially the wind energy and will adapt the solar energy on a grander scale. Wind, solar, hydrogen and fuel cells and hydropower are the most important forms of renewable energy.

Chapter 9 discusses India's policy direction and achievements in terms of renewable energy. The future thrust is also discussed at some length. Tarrif setting, schemes like the 'Green Rating for Indian Habitat' (GRIHA) are discussed.

Chapter 10 discusses the viability, the penetration, opportunities and challenges to the solar energy in India and worldwide. The relevance of solar energy for household applications and the appropriate financing strategies are emphasised in this chapter. The role of the Jawaharlal Nehru National Solar Mission in fostering the development, institution building for R&D activities and training in the arena of solar energy are given special place in this chapter. The chapter also briefly discusses about the wind energy scenario and R&D activities.

Chapter 11 discusses the importance of project management in India's power sector development. Project management principles will be discussed, used while analysing India's power sector strategy blueprint, and while analysing discussing the project performance of the major public sector company *viz.*, the National Thermal Power Corporation (NTPC) Ltd.

Chapter 12 presents the conclusions and recommendations of this book.

2 Energy Security

Framework, Pathway and Challenges for India

Introduction

Eighty-five per cent of world's energy requirements are met by fossil fuels *viz.*, oil, coal and natural gas respectively. World's energy resources are finite. Oil is expected to run out in about 40 years, natural gas in about 60 years and coal in about 200 years respectively. With the current level of technology, renewable energy cannot be made the viable substitute to fossil fuel-based energy. Due to the paucity of endowments of water supply and finity of land, renewable energy is not an economic substitute for fossil fuel-based energy in the near future. Biofuels have adverse impact on the availability of food grains causing heavy financial burden on the poor and imposes additional heavy burden of food subsidy on the governments and rationing on the people.

Hence, government should ensure energy security to India through a strategy recognising the indispensability of fossil fuel-based energy during the foreseeable future *viz.*, 30/40 years. India is the eleventh largest energy producer and sixth largest energy consumer in the world. Fossil fuels *viz.*, oil, coal and natural gas are important sources of primary energy consumption for India. Coal contributes to 54 per cent, oil to 34 per cent and natural gas, nuclear and other renewable energy sources to 12 per cent respectively. To a large extent, India depends on imports of fossil fuels (see India Briefing by USEIA, 2007).

Oil and Natural Gas

India has about 5.6 billion barrels of total proven oil reserves *viz.*, 2.4 per cent of the world crude oil reserves of over 2,300 billion barrels. The probable and proven reserves would be up to 11 billion barrels. Most of

India's oil is produced from onshore fields in Assam and offshore fields off the Mumbai coast. India is the 25th largest oil producer accounting for 1 per cent of global oil production and sixth largest consumer of oil accounting for 2.9 per cent of world's oil consumption.

India produces over 8,50,000 barrels of oil per day and consumes over 2,700,000 barrels of oil per day. India thus imports over 65 per cent of oil (see Appendix A-2.1 of this paper and USEIA, 2007). The most important source of India's oil imports is the Middle East. Saudi Arabia is the largest oil exporter to India and Qatar and Oman are the largest natural gas exporters to India. India is attempting to diversify the sources of oil and natural gas imports by including Sudan, Nigeria, Syria, Vietnam, Russia and Myanmar etc.

However, it is understood that this policy of diversification brings India into direct competition with China leading to rapid deterioration of non-Middle-Eastern oil resources. As a consequence, this would most likely make India even more dependent on Middle Eastern oil resources from the year 2020. This would no longer help India influence the price level of oil and natural gas and would subject India to scramble *vis-à-vis* the more confident and aggressive China. China has better bargaining power *vis-à-vis* India due to the ability to supply sophisticated arms and enter into technological cooperation in sensitive areas *viz.*, missile development, space application etc (see *http://iags.org/n0121043.htm*)

Turning to natural gas, we understand that India holds the reserves of over 29.32 trillion cubic feet of natural gas which is about 0.5 per cent of world's reserves. India consumes about 1.2 trillion cubic feet of natural gas per year. India imports about half of its natural gas requirements. The share of natural gas consumption is growing very fast compared to coal. India's coal consumption is growing at an annual rate of about 2.2 per cent while oil and natural gas consumption is growing at the rate of 4 per cent and 4.8 per cent respectively. This percentage growth will decline for the coal and increase for natural gas. Natural gas, alongside oil is becoming the predominant source of total primary energy consumption (TPEP) for India (see Appendix A-2.1 and (USEIA, 2007).

Coal

India is the third largest producer and consumer of coal after USA and China.

India has about 10 per cent of total world's coal reserves. Most of the domestic reserves contain relatively high amount of ash and bitumine. Most coal mines are located in Bihar, West Bengal and Madhya Pradesh states. Given the current level of production and consumption, India's coal reserves would last for more than 200 years.

India produces about 8.5 per cent of world's coal output and consumes about 9 per cent of world's coal output. About half of India's energy requirements are met by coal and 70 per cent of India's electricity is produced using coal. About 23 million tonnes *viz.*, about 5 per cent of India's coal requirements are imported. Primarily, the imports are from Australia, New Zealand etc. This is high quality *viz.*, coking coal being used in the steel industry. India only has 5 per cent of total coal reserves as coking coal (See USEIA, 2007 and *http://www.indiaenergyportal.org/overview_detail.php)*

Electricity

India is the sixth largest producer of electricity accounting for 4 per cent of production and 3.5 per cent of consumption. India's electricity production is likely to grow by 8 to 10 per cent per year into the decade of 2020-2030. (See *http://indiaenergyportal.org/overview_detail.php and http://www.cslforum.org/india.htm*)

Nuclear Energy

Under the assumption that the 440 nuclear reactors in the world and 232 proposed nuclear reactors in the world produce 556 gigawatt hours of electricity per year, 2 per cent of world's energy would be derived from nuclear power. This would amount to 16 per cent of world's electricity consumption. With regard to nuclear power, we have to take the long-term perspective *viz.*, 40 to 50 years. To achieve complete energy independence, stage three nuclear reactors have to be fully operationalised.

First stage pressurised heavy water reactors use the finite uranium-235 and uranium-238. Derived from uranium-232, U-238 helps the closing of nuclear fuel cycle multiplying the times of use of the same fuel. Second stage fast breeder reactors uses the less finite plutonium-239 using uranium-238 as the blanket. This alone could more than tripple the total energy production of India to 420 gigawatts. Third stage breeder reactors use thorium-232 potentially helping India's complete energy independence. India has world's largest thorium reserves (see *http://www.dae.gov.in/iaea/ak-paris0305.doc*).

Scheme

Section II of the chapter discusses the context, framework and the pathways to energy security. Section III discusses the sourcing and strategy towards fossil fuel security. The section discusses the sources of fossil fuels, the long-term unreliability of the sources given the unfavourable security relationship of India with neighbours like Pakistan and Bangladesh within the broader framework of global Islamic fundamentalism and the strategic disadvantage of India manifested in its inability to set the prices due to domination of multinational oil companies backed up by the powerful Anglo-Saxon alliance and attempts to and limitations of diversification of imports. The section goes on to discuss the nuclear and renewable energy scenario in India and potential limitations in the technological angle. Section IV discusses the international issues in energy security.

Section V discusses India's energy security challenges in the light of geopolitical disadvantage, technological backwardness and fluctuation in energy prices causing difficulties towards planning for renewable energy. Attempts to diversify away from the non-Middle Eastern sources, potential competition from China and rapid deterioration of non-Middle Eastern sources of oil and natural gas consequently leading to further dependence on Middle Eastern sources is a grave concern for India.

Section VI concludes and recommends that a durable energy and economic security partnership should be established with a reliable partner *viz.*, Russia, given the time tested defence and scientific relationship. This partnership could foster strategic business initiatives weaving the web of mutual economic dependence. China could be considered as a strategic partner due to large-scale trade relationship with India. India's energy security in sea lanes and pipelines could best serve China's economic and security interests.

Energy Security

Energy Adequacy

Energy security depends on the existence of adequate resource base. Resource base is the sum of reserves and resources. Reserves comprise the sum total of currently known and economically recoverable hydrocarbon deposits, natural uranium and thorium. With advances in technology, geophysics and reduction in exploration and development costs, many of today's resources are likely to become reserves.

Adequacy of Crude Oil

Over the past 20 years, proven oil reserves have increased steadily mainly in the existing fields on account of more accurate knowledge, increased productivity and advances in technology. Seismic (three- and four-dimensional) exploration led to marked improvement in the drilling techniques corresponding to horizontal and offshore drilling. This led to a significant improvement in the recovery factor from 30 per cent to 40-50 per cent.

Complimentarily, unconventional oil reserves *viz.*, the heavy oil, the tar sands, oil shale deposits etc., are yet to be exploited due to economic reasons. For example, 1.2 trillion barrels of heavy oil are found in the Orinoco oil belt in Venezuela. And the tar sands of Canada and oil shale deposits of the Russian Federation may contain 300 billion barrels of oil.

Oil scarcity is not likely to occur until 2020 and to meet the probable scarcity thereafter; unconventional oil reserves would be utilised. One good news pertaining to the oil is that, the importance of oil is being reduced in the global energy scenario.

Between 1973 and 2010, the share of oil in the fossil fuel consumption and imports declined from over 46 per cent and 50 per cent to about 40 per cent and 42 per cent respectively.

Natural Gas

The cumulative production of natural gas until 1993 amounted to 1,750 trillion cubic feet (1,722 exajoules, or 41 Gtoe) which was equivalent to 17.1 per cent of the US Geological Survey's estimate of conventional gas reserves. Natural gas consumption is projected to grow by 2.6 per cent a year, primarily on account of electricity production in non-OPEC countries. At this rate, natural gas reserves can cater to the global needs for the next 60 years or so.

Coal

Coal is the world's most abundant fossil fuel, with reserves estimated at almost 1,000 billion tonnes which equal 27,300 exajoules, or 650,000 million tonnes of oil equivalent (Mtoe) (WEC, 1998). At the present rate of production, these reserves should last for more than 220 years.

Besides, coal is more evenly distributed across the world, cheap and cost-effective having greater scope for emission reduction through the development and use of clean coal technologies. The future of coal demand is hard to predict due to the restrictions imposed by environmental policies. However, gasification and liquefaction of coal is contributing to the prolongation of oil and natural gas reserves. Incidentally, the combined cycle coal gasification power plants have achieved the energy conversion efficiency of up to 70 per cent, almost double the rate of energy conversion efficiency in the coal powered power plants.

Nuclear Energy

In 1998, global nuclear energy production amounted to 2,350 terawatt hours of electricity, replacing 620 Mtoe of other fuels. Uranium utilisation was to the extent of 63,700 tonnes in 1997. Up to 17 million tonnes of recoverable reserves of uranium, the discovery of new reserves of low grade uranium and vast deposits of thorium give a reassuring picture concerning the nuclear power. Nuclear reprocessing also help multiply the lifetime of a nuclear power plant by at least 30 times to 2,000 years.

Renewable Energy

Renewable sources of energy, especially hydropower, biomass, wind power and geothermal energy account for a growing share of world's energy consumption. Hydropower and biomass together account for 15 per cent of energy consumption. Hydro electricity contributes around 2,500 terawatt hours of electricity a year. It replaces almost 675 Mtoe of fossil fuels a year.

Technically exploitable hydro resources could produce more than 14,000 terawatt hours of electricity annually (WEC, 1998). Environmental and economic reasons hinder the exploitation of most of these resources. Hydropower growth potential is very high *vis-à-vis* the growth potential of the demand for energy due to cost-effectiveness, eco-friendly nature and the conversion efficiency of almost 100 per cent.

Biomass is the chief source of renewable energy consumed by over a third of global population—mainly in developing countries. It comprises fuel wood, crop residue and dung. Although theoretically sustainable, the energy conversion efficiency is very low *viz.*, hardly 10 per cent. Besides, the rate of deforestation is higher than the rate of afforestation reducing the carbon intake and increasing global warming and ecological disasters.

Biomass energy technology is advancing rapidly. Besides direct combustion, gasification, fermentation and anaerobic digestion techniques are all increasing the potential of biomass as a sustainable source of energy. Besides biomass, wind power could potentially contribute to over 10 per cent of world's energy requirement and geothermal and gas hydrates have the future after 2030-2050.

Supply-Side Security of Energy

As noted in chapter 1, most of oil resources are concentrated in few regions and even natural gas is not widely available. Competition between the developing countries—particularly between the bigger economies *viz.*, India, China etc., put pressure on oil reserves and on the price of oil. Besides, the share developing countries in the use of oil and natural gas is increasing fast. From 70 per cent in 1980, the share of OECD in world oil consumption reduced to just over 50 per cent.

Technological advances in oil exploration and recovery could at best prolong the supplies by few decades. Besides, the impossibility of drilling for oil and natural gas in large population centres and environmental damage control hinders the full exploitation of oil. Insecurity of supply lines (geopolitical) is prevalent and perhaps increasing due to potential terrorist attacks on the sea lanes and pipelines.

Geopolitical security to sea lanes and pipelines is being enhanced in OECD countries through upstream and downstream partnerships between the companies of oil and natural gas producing and oil and natural gas consuming countries. The partnerships also extend to the ownership and management of pipelines and security of sea lanes *viz.*, the Indian Ocean Rim-Association for Regional Cooperation (IOR-ARC) watching over the Malacca Straits and the Gulf of Aden under the leadership of India and the United States of America with contributions from China and Southeast Asian countries.

The risk of supply disruptions due to serious accidents of tankers and pipelines and due to the terrorist attacks on the oil and natural gas infrastructure is mitigated using long-term contracts of production and delivery between the importing and exporting countries. Since fossil fuel-based energy supplies are finite involving potential serious geopolitical risks, and biomass-based energy is not fully assuring, long-term energy security measures are imperative.

Diversifying the sources of energy supply and forms of energy use, international cooperation and agreements among energy importing countries and between importing countries and exporting countries involved in or not involved in production and among governments and companies involved in investment and transfer of technology to developing countries play a key role in ensuring long-term energy security.

Roadmap to Energy Security

Energy security can be ensured by following the policy of local adequacy: especially for large countries like India. Towards this endeavour, abundant and diverse forms of indigeneous energy resources should be harnessed to enhance energy security. Transportation network, grids and pipelines should be used to import energy resources to tide over local shortages.

Strategic stocks of energy resources should be held in strategic locations to tide over the transient interruptions, shortages and unforeseen surge in demand. Technological and financial resources and the knowledge and skills should be developed and used for the development and harnessing of the renewable energy resources to fully or partly meet the national energy requirements.

Due to the impending scarcity of fossil fuels and severe price fluctuations, energy conservation and efficiency measures should be used to reduce energy intensity and in turn to reduce dependence on the imports of fossil fuels.

Energy Security: An Analytical Framework

Energy security has to be ensured on the demand as well as on the supply side.

Demand-Side Factors

1. *Understanding the demand structure*: Four user groups are to be studied for this purpose: household, agricultural, commercial and industrial users, their demand pattern, the contribution of demographic trends etc., must all be studied.
2. *Ability and willingness of the users to pay*: The project planning stage should encompass the ability and willingness to pay so that right quantity can be sold at right price. This also substantially mitigates the risk of theft and loss.
3. *Linking energy services with user's income status*: People's income status determines their purchasing power. The present and future income status of the people should be taken into consideration while determining the provision of the quantity and quality of energy services.

Supply-Side Factors

1. *Macroeconomic indicators*: Sustained provision of energy services is possible if favourable macroeconomic factors exist. Energy importing countries need a favourable and stable price. Poorest countries like Rwanda etc., experienced a doubling of oil prices several times during 1990s and in 2007 in a span of over a year. This severely affected the purchasing power of the poor population.

   Incidentally, kerosene and mass transportation became dearer by between 70 per cent and 100 per cent respectively. This is a severe pinch on the bulk of population *viz.*, about 70 per cent who happen to be poor. Subsidy burden in countries like India increases pertaining to LPG and kerosene. This severely affects the balance of payment, foreign exchange reserves and the value of currency along with the adverse effects on the Terms of Trade.

2. *Geopolitical strategies*: The geographic concentration of hydrocarbon resources leads the countries to enter into bilateral or regional agreements towards ensuring stable oil and natural gas supplies. Geopolitical relationships are characterised by mutually beneficial arrangements. For example, China trades arms and diplomatic benefits for oil and natural gas far edging out India.
3. *Strengthening institutional capacity*: Developing countries can be categorised as net importers or net exporters of energy. Net exporting countries such as Chad, Algeria, Nigeria and Angola must have corporate and institutional structures in place to sustain gains from their energy exports, especially the gains from recent oil price increases. Only a robust institutional setup can channelise the gains from oil and gas exports into projects of broader economic development. Net importers can use their geographical and technological capabilities to generate and sustain revenues from refining and processing of hydrocarbons.
4. *Strengthening financial resources*: The total investment need in the energy infrastructure for the next 25 years is projected to be $16 trillion. To provide optimum quality and quantity energy services, developing countries need to mobilise resources in the domestic and international capital markets. Investment flows to countries practicing good governance and taking anti-corruption initiatives.
5. *Environmental impact mitigation*: The adverse environmental impacts due to the excess use of hydrocarbon energy results in smog and carbon emission adversely affecting agriculture and health status of the population. Countries wishing to sustain economic development must mitigate the adverse impact on the environment either on their own or through international cooperation.
6. *Natural resource endowments and their feasibility*: Understanding about the current and future resource endowments, feasibility of domestic and international technology in development and production of such resources should be inbuilt in the energy policy of the countries.

Energy Resource Feasibility

Mapping the energy resources, feasibility of developing and using those resources and trading in those resources should be analysed using the framework based on the above discussed demand- and supply-side factors and macroeconomic indicators which is also given below in the Box 2.1.

Box 2.1

Country Optimisation Model

Ability of a country to optimise its energy resource portfolio and supply of energy services for the desired level of services that sustain economic growth and poverty reduction.

Demand-side Analysis	Supply-side Analysis
Understanding the demand structure – Factors affecting consumption of energy services for each user group (industrial, agricultural, community & households) – Current consumption levels – Current expenditures on energy services – Impacts of service disruption – Securing demand for energy services – Demand variations due to seasonal changes, transitioning economies, etc. – Ability & willingness of users to pay for services for energy resources (Adapted from TERI Volume on Energy Security, 2008).	– Macroeconomic factors – Balance of payments – Budgetary impacts of subsidies – Foreign exchange reserves – Financial factors – Securing incomes among user groups to pay for services – Ability of a country to attract domestic & international investment – Environmental sustainability impact of expanding energy services – Natural resource & technological endowment – Geopolitical supply agreements – Institutional capacity

contd...

....*contd...*

This framework can be translated into specific tasks which as outlined below:

Energy Security Subcategories	*Tasks*
Macroeconomic factors	Measuring the import dependence of each country to understand its exposure to fuel price volatility
Demand & demographics	Understanding the demand function of each user group according to consumption patterns and the ability to pay for energy services
Geopolitics and documenting existing geopolitical supply agreements and institutional capacity	Analysing changes in demand when economies are undergoing restructuring
	Corporate structure required to sustain energy imports or gains from energy exports
Financial feasibility	Documenting the ability of a country to attract domestic and international investment to expand energy infrastructure and service delivery
Environmental feasibility	Understanding the impacts of expanding access on the environment and the role of technology (e.g. improved devices such as cooking stoves, renewable energy generation) in mitigating these impacts
Natural resource endowment & technological feasibility	Mapping the energy resource endowment of each country and the economics of developing those resources

Emerging Issues and Challenges to Energy Security

Energy security is being investigated at global, regional, national and at the consumer levels. Ensuring global resource adequacy, ensuring regional networking and trade, ensuring national energy supply security, and ensuring the satisfaction of consumer demand are essential components of the energy security strategy. At the national level, energy security depends on the availability of all the energy commodities at all times.

Usually, governments take care of domestic exploration and production, maintenance of strategic energy reserves and imports. However, increasing autonomy is being given to the markets in these areas along with the determination of prices. At the consumer level, market players should be made responsible for price stability and steady supply of energy commodities. Formation of cartels among the companies and power utilities should be discouraged and prevented.

Supply security implies resource security as well as the security of actual delivery of the energy resources and commodities *viz.*, the coal, oil, natural gas, electricity, petrol etc. Besides the quality and adequacy of energy commodities, the security of the infrastructure from the risks of vintage, vandalisation and terrorist attacks needs to be ensured.

The long-term energy security hinges on the development of renewable energy resources and technologies to harness these resources and convert them into commodities. Promotion of energy conservation measures and the reduction of energy intensity prolong the life span of the energy resources: especially the fossil fuels-based coal, oil and natural gas.

Energy intensity shows the volume of energy necessary to produce a unit of GDP. In USA, energy intensity in the form of primary energy consumption reduced by about 80 per cent in the past 200 years and by about 20 per cent in the past 20 years. Energy intensity in middle income developing countries is three times that of the industrialised countries. Energy intensity in low income developing countries is increasing due to the setting up of energy intensive industries and the provision of energy subsidies and the use of outdated high energy consuming plant and equipment.

The increased use of coal along with the outdated plant and equipment not only harms the environment, but also contributes to the wastage of resource. Coal's energy conversion efficiency is only about 25-35 per cent compared to over 50 per cent in oil and natural gas.

Developing countries are taking off in their economic growth. Once the take off stage is crossed, countries could focus on better technologies promoting energy efficiency. Phasing out of subsidies and the improvement of the environment awareness does contribute to the reduction of energy intensity.

On the technological front, the combined cycle gas turbines have the potential to improve the energy conversion to 70 per cent far above the global average of about 31 per cent in all the electricity producing plants. The energy conversion in thermal power technology has improved to over 40 per cent. On the whole, electricity is contributing to the reduction in energy intensity by over 50 per cent.

In the industrialised countries, energy conservation became important in the aftermath of the 1973 and 1979 oil crisis. The oil crisis of 1970s was sparked by the dependency of the OECD countries on foreign oil and the inadequacy of domestic oil resources. In 1950 oil accounted for 29 per cent of total energy used in OECD countries of which, 29 per cent or 2 million barrels were imported daily.

In stark contrast, by 1973, 53 per cent of the total energy use was through oil and 67 per cent *viz.*, 26 million barrels daily were imported.

The price reduction was modest and again increased sharply by 1979 causing the major oil shock. Price of all the energy forms rose substantially in OECD countries and more so in the United States contributing substantially to the inflationary thrust. The positive effect of the oil shocks manifested in the form of the growth of energy conservation technologies.

Between 1973 and 1986, the energy demand rose only by 2 per cent per year whereas, the GDP rose by 2.9 per cent per year which shows the energy intensity in OECD countries of about 0.66. Currently, India is also manifesting the reduction in energy intensity although it is up to 0.9. However, the marginal rate of energy intensity *viz.*, the ratio of the growth rate in energy consumption to the ratio of the growth rate in GDP is very low *viz.*, below 0.75, which is a highly encouraging scenario concerning India. In future, the marginal intensity would reduce further pulling down the average intensity.

Sourcing and Strategy towards Fossil Fuel Security for India

Oil and Natural Gas

Source Profile

About 65 per cent of India's oil import is from the Middle East. The dependence is growing rapidly. *World Energy Outlook* (Interntional Energy Agency), projects that India's dependence on oil imports will grow to 91.6 per cent by the year 2020 (see *http://www.iags.org/n0121043.htm).* As noted in the introductory section, India is trying to diversify the countries and regions of oil import. Towards this effect, India adopts the policy of direct engagement with the Governments of oil exporting countries rather than with the multinational companies like Shell etc.

In Africa, Sudan and Nigeria are shaping into important partners. In Sudan, India has invested $750 million in exploration, development and production of oil. India purchases about 44 million barrels of crude oil per year from Nigeria on a long-term basis. Turning to Asia, Russia is becoming important partner of India. (see *http://www.iags.org/n0121043.htm)* Indo-Russian partnership assumes added importance due to the growing need for natural gas in India.

India imports almost half of the natural gas requirements. The most economic route to import natural gas is assumed to be through the proposed gas pipelines. The three proposed pipelines *viz.*, the Iran-Pakistan-India gas pipeline, the Turkmenistan-Afghanistan-Pakistan-India (Trans-Afghan) gas pipeline and the Myanmar-Bangladesh-India pipelines all have security risks, geopolitical risks and market risks.

Even the Myanmar-India pipeline bypassing Bangladesh has market risk covertly backed up by geopolitical risk. Hence, Indo-Russian partnership is envisaged as a platform for investment in natural gas fields *viz.*, the Sakhalin in Russia and elsewhere. The recent Indo-Russian partnership to explore and develop oil and natural gas fields in Myanmar could break the deadlock and could favour India's quest for energy security.

India's ONGC holds 15 per cent stake *viz.*, 1.7 billion dollar stake in the Sakhalin-1 oil and gas field. The field has the proven oil reserves of 2.3

billion barrels of crude oil and over 17 trillion cubic feet of natural gas reserves respectively. Russia and India are negotiating towards ONGC's stake in Sakhalin-3 oil and natural gas project. (see *http://www.jamestown.org/single/?no_cache=1&tx_ttnews per cent5Btt_news per cent5D=33168)*

If on an average, 2,00,000 barrels of oil is produced each day, the Sakhalin-1 partnership will last over 30 years *viz.*, until the year 2036. If the Sakhalin-3 partnership materialises, Russia will become important source of oil and natural gas import for India after the Middle East. Within the Middle East, India wishes to diversify away from Saudi Arabia for oil and Qatar and Oman for natural gas.

The Iran-Pakistan-India gas pipeline has to be understood in the light of diversification. India and Iran have a cooperative relationship with regard to oil and natural gas. Indian Oil Corporation (IOC) and Petropars of Iran are cooperating on a $3 billion project to develop a gas field and setting up of a gas liquefaction plant. This should be seen in the light of the proposed India-Iran gas pipeline to import gas into India, the pipeline will supply 55 billion cubic metres of gas to India per year. The construction of the pipeline is likely to be undertaken between 2009 and 2012.

Quest for Strategic Ties

Caspian region including Russia is fast becoming important source of oil and natural gas to India. The region comprises Azerbaijan, Iran, Turkmenistan, Kazakhstan and Russia. India has been cultivating good relationship with Caspian countries right from the Soviet era. Especially, Russia and India have similar stand with respect to Islamic terrorism and local separatist movements. Both India and Russia are the victims of terrorism and separatist movements.

India is forging strategic relationship with Tajikistan, Turkmenistan and Kazakhstan by providing financial aid and establishing military bases. Incidentally, Tajikistan hosts India's first overseas air base. Iran is also becoming strategic partner to India. While it is the opposition to Taliban and the quest for closer economic ties with India that forge Iran closer to India, opposition to Islamic fundamentalism and the desire to be part of the security compact between India and Russia forge Turkmenistan and

Kazakhstan closer to India. Incidentally the Farkhor air base in Tajikistan operated by India is only 2 km away from the Afghanistan border.

India is trying to forge long-term ties with Myanmar and Vietnam to explore, develop and produce oil and natural gas from the offshore fields in Bay of Bengal and South China Sea. India's strategic quest has to be seen in the light of global energy scenario.

Global Energy Scenario

Global energy scenario reflects the complicated interplay between nations and companies. The chief importers of energy in the world include USA, Western Europe, China and India while the chief exporters include The Middle East and the Caspian regions of the world. Energy resources are scarce. Different countries and groups of countries adopt different approaches to secure energy supplies.

Anglo-Saxon countries *viz.*, the USA and the UK and few Western European countries *viz.*, France, Germany, Italy and Belgium tend to take the route of domination of oil producing regions and the subsequent creation and transfer of property rights while China and India tend to prefer to enter into direct legal contract with energy producing countries and companies.

The Anglo-Saxon and Western European countries tend to favour private transnational oil corporations while India and China tend to favour public sector corporations. Caspian littoral nations *viz.*, Azerbaijan, Iran, Kazakhstan, Russia and Turkmenistan own about 21.6 per cent of total proven oil reserves and 45.6 per cent of proven natural gas reserves. (see Amineha and Howelingb, 2007).

Nuclear and Renewable Energy

Using the current domestic reserves of uranium, India's nuclear power reactors produce 3 per cent of India's electricity. To raise the share of nuclear power to 15-20 per cent within a decade viz., by the year 2020, India must import uranium from the countries together known as the 'Nuclear Suppliers Group (NSG).' This is with the understanding that complete energy independence through the three stage nuclear programme is a distant scenario.

The NSG comprises 45 countries led by the USA. Australia, Austria, Canada are important countries in the group. To be eligible for nuclear technology assistance and uranium imports, the country should be the signatory to the Nuclear Non-Proliferation Treaty (NPT). However, the recent Indo-US Nuclear Accord exempts India and makes her eligible for import of uranium. However, India has to subject the so-called civilian nuclear reactors to International Atomic Energy Agency (IAEA) inspection to ensure the usage of the reactors for civilian purpose only.

There is very important concern that India has to sacrifice technological independence to be eligible for uranium imports which in the long run could jeopardise energy security. Presently, Indian nuclear power reactors operate at 46 per cent capacity producing 4,100 megawatts of power. Imported uranium could increase the capacity to over 60 per cent immediately (see *http://www.dae.gov.in/iaea/ak-paris0305.doc.)*

Turning to renewable energy, we understand from the Ministry of New and Renewable Energy (MNRE) that renewable energy is small but significant component of India's energy production. The total installed capacity is 13,728 megawatt or about 12 per cent of total produced energy. Wind energy dominates with over 9,521 megawatts followed by the small hydro projects producing over 2,200 megawatts. Wind energy still exhibits the fastest growth with about 8 per cent. India is the fifth largest producer of wind energy. However, the favourable locations are mostly confined to southeastern coast. (see *http://mnes.nic.in/)*.

International Energy Security Concerns

International Energy Security Analysis

International energy security analysis has multiple facets. Some of them are discussed below.

a. The Impact of Price Levels and Price Changes on the Energy Consumption

Countries with highly successful export oriented economies are often among the largest importers of energy resources *viz.*, the oil and natural gas.

Countries like Japan, South Korea, etc., enjoyed very strong economic growth rates and were highly export oriented. However, they are heavily dependant on imported sources of oil, natural gas and coal. Often, the fluctuations in the international energy prices and at times their rapid escalation proved costly for these countries. Hence, the focus on the development and use of alternative energy resources and the energy security frameworks became imperative.

Energy security frameworks should yield the highest possible net benefits to the country. The frameworks tend to have one or two or three focus areas. Financial measures, technical measures and policy measures are the three prime focus areas.

Analytical techniques are not able to properly capture the cost/benefit impact of the trade-off between the focus areas.

Along with physical factors, economic factors tend to play a vital role in enabling the availability of energy resources. Physical disruption to the supply of oil in 1979 did not affect all the importing countries uniformly. Countries which imposed price controls such as the USA suffered greater physical disruption compared to countries which did not such as Japan. Japan did not hesitate to pay higher price for oil.

Similar analysis can be extended to the choice of other types of fuels and for understanding the relative importance of each form of fuel in the economy subject to the physical, financial and technological constraints. Together, these constraints make up the economic constraint which in turn can be the basis for understanding the relative economic cost of each fuel. Arrival at the economic cost requires the incorporation of specific risks related to the sources of the supply of a particular fuel and the systemic risk pertaining to the concerned fuel market itself.

b. Geopolitical Risks or Quantifying Supply Diversification

Herfindahl measure of market concentration is used to quantify the supply diversification or dependency. The Herfindahl index is the sum of the squares of market fractions controlled by individual firms. In our case, supplier countries and the types of fuel come under the perview of this definition.

The simple Herfindahl—or diversification index is presented as:

2 H =å xi i

Where xi is the fraction of total supply from source 'i'.

Using this logic, we can discern that if there is only one supplier, the index is 1. If there are 10 suppliers with equal market shares, the index is 0.1.

If there is one dominant supplier with 90 per cent market share and one other supplier with 10 per cent, the index is 0.82.

A lower index implies greater diversification of supply. The reciprocal (1/H) of the Herfindahl index can be viewed as the number of suppliers with equal market shares.

To use this index, the risks associated with other types of fuels must not correlate with the risks associated with the fuel type under analysis. Fuel diversification can be computed across companies, countries and country groups using this index.

In the case of India, price risks across the domestic companies and foreign companies can be computed for oil and natural gas using this approach.

Similarly, geopolitical supply risk can be computed across the countries or the country groups using the indicators of security threats.

Market shares of the source of suppliers can be compared with the incidents of supply disruption due to price and non-price factors. For example, comparison of the country groups is undertaken pitching the OPEC nations *versus* the nations with which India has signed long-term supply contracts and partnership deals.

OPEC countries are led by the Saudi Arabia and the Middle East while the countries with which India has signed long-term supply contracts include Russia, Sudan, Angola etc. The subsequent subsections discuss India's energy security concerns in the light of economic and geopolitical supply risks. But in this section, we confine ourselves to the international energy security concerns and the analytical approaches to understanding of and mitigation of economic, geopolitical and environmental risks.

c. Environmental and Fuel Diversification

Five types of fuels dominate the global energy supply at present. They are: [1] oil, [2] gas, [3] coal, [4] nuclear and [5] hydro.

If a nation uses all the types of fuel to derive identical proportion of energy (assuming the fuel types to be ultimately interchangeable in use), the diversification index would be:

1 H fuel ( )2 020 = 5 = . 5

The index of the fuel types in use should be 0.2 for each type of fuel if the proportion of use is equal. However, it is not equal in reality.

The BP *Annual Statistical Review of World Energy*, 2006 states that the world energy consumption in 1996 was 8,380 million tonnes of oil-equivalent. The proportion of each fuel consumed was 0.395, 0.235, 0.269, 0.074 and 0.026 respectively.

Trends suggest that the proportion of oil, gas and coal has to come down and that of nuclear power would increase substantially along with the renewable sources of energy *viz.*, wind power and solar power in the long run.

By squaring of the proportions, and by adding those squares we find the average global diversification index to be 0.290. Even if hydropower and nuclear fuel are not accounted for, the index would be 0.294. The index for Asia-Pacific countries without China is the highest and stands at 0.367 and for Japan it is 0.358. Among the Western OECD countries, US' and Great Britain's indices approximate the world average and Canada's index at 0.256 is the most diversified.

Trade-off to Achieve Greater Energy Supply Security

Energy diversification has to be achieved by adapting either of the below approaches:

a. Improving the energy mix.
b. Establishing energy intensive industries in the energy self-sufficient countries and less energy intensive industries and service enterprises in the energy deficit countries.

c. Comparing the cost of fossil fuel waste with the cost of nuclear waste or the waste generated from the alternate fuel.

d. Acceptance and assimilation of the environmental impact eminating from the choice of industries and energy mix.

The global implications of the energy mix would either reflect the global shift or redistribution of the energy mix. For example, energy intensive industries would be set up in the developing countries so that developed countries could reduce their carbon footprint. The energy import bill, or the negative environmental impact due to the running of the energy intensive industries or a combination of both would adversely affect the developing countries. Hence, a shift towards a sustainable energy mix and cost reduction should be universal. Developed countries and countries with advanced technologies. Pertaining to renewable energy should share them with developed countries. Setting up of more energy intensive industries in the energy sufficient countries cannot be the long-term solution due to the environmental considerations.

Supply or fuel diversification indices should not reflect geographical biases. Harnessing and the transportation of the domestic fuels may at times be more difficult and risky compared to importing the said fuels. Domestic sources may be afflicted by strikes, accidents arising out of poor safety standards, poor transportation facilities and intense environmental agitations due to new or already understood environmental problems. Under these circumstances, import of the fuels may be more feasible and tenable physically as well as financially.

The concept of supply diversification can be extended further to supplier diversification. Taking oil, we understand that the dominant supplier is the OPEC. Apart from OPEC, Russia, Angola and other countries are also important sources of supply. There is a maximum potential supplier diversification for each country or group of countries. For example, in the case of the OPEC countries, shelving or cutting the production, the diversification gets reduced. Even in the presence of the diversification, individual countries or groups of countries are constrained to opt for few suppliers on a long-term basis due to financial or geopolitical reasons.

For example, global supplier diversification using the oil production data is 0.6 and using the oil export data including the OPEC data is 0.20. However, the US and the Western Europe import more oil from North and West Africa along with from Venezuela and Columbia. Hence, the diversification is 0.242 for these countries. For Japan and for the Asia-Pacific region, the index is 0.604 and 0.628 respectively thanks to the excessive dependence on the Middle Eastern oil. Same is the case for India which imports about 2/3rd of oil from the Middle Eastern region which is further expected to increase to 90 per cent in 15-20 years as we shall discuss in the subsequent subsections.

There is a trade-off between the choice of the suppliers and marginal economic considerations. For example, Japan uses vast amount of cheaper and heavier sour crude for electricity generation. The additional cost of diversifying the suppliers would be 10 per cent in the case of several countries. Supplier diversification can reduce specific risks related to the supplier and supply roots. However, systemic or the market risks cannot be reduced by the supplier diversification.

The general level of fuel prices and the price movements depend on the availability of the fuel on the 'spot' and 'futures' markets etc., subject to the technological and decision constraints as we discuss in Chapter 5. The answer to high market risks would be in the adaption of renewable energy technologies that are sustainable. Solar energy and to some extent Wind energy and biomass energy offer the answer to sustainability.

Overcoming Systemic or Market Risk

Market risk occurs in the event of uncertain physical supply in the market. The uncertainty arises often due to war and accidents to the tankers and pipelines. To overcome the market risk, physical supply security should be ensured. In case of disruption to the production and supply from Saudi Arabia, India should be able to source the oil and gas from countries like Russia etc. In case of supply cutbacks, other sources should be able to replenish the vacuum.

Practical experience based on the modern portfolio theory suggests that oil or any energy commodity should be sourced from as many suppliers as possible so that the supply variance can be minimised. During the 1991 Gulf war, countries which exclusively or predominantly depended on Iraq and Kuwait for oil suffered most unlike the countries which diversified their supply sources. The risk of depending on one or few suppliers was thrice the risk to be faced after the maximum potential diversification.

It is most likely that the psychological impact of physical supply disruption gets escalated in the aftermath of the physical supply disruption. Countries tend to increase their inventories rapidly in the event of physical supply disruption. This causes a rapid price escalation. A systematic policy should be in place to mitigate the systemic risk.

Apart from maximum possible diversification of supply sources, focus should be on building the strategic reserves of petroleum and natural gas. At present, the trend is to build the strategic reserves for three months scaleable to six months' worth supplies.

Does Domestic Production of Fuels Minimise the Systemic Risk?

There is no definite answer to this question. The notions, concerns about and the incidents of strikes, accidents and environmental legislations along with the need to honor the international agreements aimed at climate change mitigation all contribute to the systemic risks pertaining to the domestic supply as well. Hence, minimisation of the supply variance using all the sources of physical supply irrespective of the nativity should be undertaken to minimise the systemic risks. Indicators of risk should be identified and quantified to undertake proper analysis and plan for risk mitigation. Monopoly of the US in the nuclear spent fuel reprocessing and supply cutbacks by the NSG that supplied uranium resulted in high demand and a faster price escalation in the aftermath of the 1974 India's nuclear test.

Approaches to Mitigate the Systemic or the Market Risk

a. *Investment*: It is possible to invest in risk minimisation. For example, there are many suppliers of a fuel *viz.*, in the upstream. However, the downstream, especially various steps of processing and especially in the nuclear power industry impede the risk mitigation efforts. Investment should be made in the right kind of company, or product or the process. Right kind of investment can make sure the discovery and production using low cost methods such as those used in Canada. Thus, direct investment tend to reduce the variance of the source and its correlation with the overall market portfolio risk.

b. *Contracting*: Reduction of variance often requires long-term contracting. Long-term contracts often promote supply stability through the medium of futures and options contract. This is especially beneficial for the import dependant countries.

c. *Inventories*: The strategic reserves are the costliest way to mitigate the systemic risk. Fuel has to be purchased at the market price to build strategic reserves. Strategic reserves contribute to a reduction in the variance. Are often considered as an additional supplier.

However, the strategic reserves should only be drawn during the emergency situations. The cost of buying and holding the reserves is much higher due to the market price paid for the fuel and due to the infrastructure to be established to hold those reserves.

Cost varies across the fuel types. For example, the cost of holding the inventory of uranium is only 3 per cent of the cost of holding oil. If low enriched uranium ready to be converted into electricity is stock piled, the cost would only be 8 per cent of the cost of holding the oil. Thus, the cost of holding oil and coal is much more than the cost of holding uranium and nuclear fuel.

India's Energy Security Challenge and Determinants of Strategy

The Challenge

India's energy security challenge has to be understood in the light of India's position in global geopolitics, technological backwardness and strategic disadvantage. First, powerful elements in the Gulf countries including in Iran sponsor terrorism. Any deterioration in relationship between India and Pakistan or any adverse activism by local fundamentalist Islamic elements in support of Palestine against Israel or in opposition to strengthening defence and aerospace ties between Israel and India is most likely to adversely affect the oil supplies or the 'Terms of Trade.' Hence, India is diversifying her oil and natural gas imports.

Second, as a consequence of diversification, India enters into direct competition with China in Africa, in South China Sea and even in Myanmar. India is forced to invest in partnership with China in Angola, Cote d'Ivoire and Sudan. There is no guarantee that Indian investments will result in oil and natural gas inflows from Myanmar and possibly even Vietnam.

China is the third largest economy in the world and is likely to grow by 8-9 per cent per year until 2030. China's demand for imported oil is growing rapidly. From a practically non-existent import of crude oil before 1993, China now imports 2.9 million out of 6.5 million barrels a day *viz.*, about 45 per cent of total requirements. These imports are likely to grow by 960 per cent in the next two decades, most probably taking the share of imports to 70 per cent of total requirements.

India has emerged as the fourth largest economy in the world and is the sixth largest energy consumer. Currently, India produces only 30 per cent of its oil and about 50 per cent of its natural gas requirements. This implies that India imports 70 per cent of oil and about 50 per cent of natural gas respectively. The percentage of import is soon likely to cross 91 per cent according to the Institute for the Analysis of Global Security (IAGS) (see *http://www.iags.org/n0121043.htm*)

This makes India 30 per cent more dependent than China. Greater the dependence on imported oil, lesser the bargaining power with oil producers. India doesn't have as much proactive diplomacy unlike China. India's policy is only to provide 'line of credit' to developing countries, especially while promoting joint venture projects. Whereas China tends to promote holistic relationship by developing the physical and human infrastructure. For example, China takes special interest in developing ports, railways and roads in Myanmar, Cambodia etc., thus earning better edge over India.

Third, India is neither part of the US led block nor Russia led security pact to have a proactive defence services backed up energy security policy that would help tide over energy price fluctuations through aggressive trade and investment policy. Hence, building up of strategic reserves, or adapting energy efficiency measures or cutting non-essential fuel subsidy seems to be the strategy of immediate relevance.

Fourth, India's assertive strategy could be counterproductive. If India uses assertive strategies to secure energy security interests, China might use the geopolitical advantage as a deterrent. Myanmar's ports could be used by Chinese navy, and railways and roadways by Chinese army in case of aggression against India. Pakistan being the transit route of India-Pakistan-Iran gas pipeline and the Trans-Afghan gas pipeline, India has the geopolitical disadvantage in this respect.

Poor security relationship between India and Pakistan and difficulty in the development of trustworthy relationship between India and Bangladesh pose hurdles in the fructification of the India-Pakistan-Iran gas pipeline, Trans-Afghan gas pipeline and Myanmar-Bangladesh-India gas pipeline.

The adverse domestic political situation in Bangladesh and the presence of anti-India militant groups in Bangladesh led to great delay in the commencement of Myanmar, Bangladesh India gas pipeline for over 12 years. Hence, Myanmar inked deal with PetroChina—a Hongkong listed company to supply gas from the offshore A1 block. What shocked India is that the deal was reached without her knowledge. China, thus, benefits through the PetroChina-Myanmar deal. This is in spite of the fact that India's ONGC and Gail hold 20 per cent and 10 per cent of the stake in A1 block respectively and the South Korean firms hold the remaining 70 per cent respectively.

The PetroChina deal and China's military activity on Bangladesh's soil are in spite of Gen Maung Aye's emphatic declaration that 'Myanmar will not allow its territory to be used for any hostile activities directed against India.' In fact, Myanmar discouraged terrorists operating in the northeastern region from using its territory even during the times when Indian media including the official All India Radio (AIR) were indulging in anti-Myanmar propaganda due to the military rule and particularly due to the arrest of Omsang Souki. (see *http://www.southasiaanalysis.org/papers2/paper197.htm* and *http://www.saag.org/common/uploaded_files/paper1681.html*)

Attempt to lay an alternate pipeline would only increase the distance by 150 km to 440 km and lead to cost escalation. India is yet to fully build confidence among the population of the northeastern states and hence the pipeline is not assuredly fully secure via Pakistan (see *http://www.southasiaanalysis.org/papers2/paper197.htm and http://www.saag.org/common/uploaded_files/paper1681.html*)

India is also yet to build confidence with African nations. While India provides some 'line of credit' to promote certain projects like railways in African countries, China is developing multifaceted relationship. Provision of scholarship to thousands of students, establishment of military relationships, development of infrastructure on a loan/grant basis etc., is the plus point for China. China has also sponsored the liberation struggles in Africa.

The adverse security relationship between India and Pakistan and threats from the fundamentalist Islamic elements has consequently resulted in the delay in the conceptualisation and construction leading to more than quadrupling of the estimated cost of the 2,775 km long pipeline to over $7.5 billion during the period 1990 and 2009. India has to spend $5.53 per British thermal unit of gas including the 60 cent transit fee payable to Pakistan.

The price is over a third higher than the prevailing market price in India. Recent Press Trust of India (PTI) report said that Iran is demanding $10 to $11 per Btu of gas and hence Pakistan is reconsidering its decision to be a partner in the project. By the year 2020, the proposed pipeline will constitute over 37 per cent of the total natural gas requirements of India, if India consumes 146 billion cubic metres of gas per year (see *http://en.wikipedia.org/wiki/Iran-Pakistan-India_gas_pipeline* and *http://www.dnaindia. com/ report.asp?newsid=1221711*)

Fifth, India has to grapple with its own problem of energy inefficiency. India's energy intensity is 0.91 which is about double the most energy efficient economies. Even China and Pakistan have lower energy intensity *viz.*, about 0.6. Thus, use of energy saving technologies is imperative for India.

Using Lord Nicholas Stern's (Stern Review, The Economics of Climate Change 2006) argument, saving energy could entail curtailing of GDP growth. If GDP growth is curtailed by about 1 per cent per year, enough resources can be found for investment in renewable energy technologies.

Else, the world could experience GDP shrinkage of about 5 per cent per year from 2050. Part of the resources could also be invested for coping with climate change.

Sixth, India and the world as a whole is facing challenge from divergence in expectations. Sudden drop in oil (crude oil) price from $147 to $36 in the later part of 2008 and during early 2009 has made plans towards renewable energy investment redundant. Several renewable energy technologies *viz.*, the wind power, co-gasification etc., will be viable only if the oil price is $70 or more per barrel. With the oil price shooting above $70 in June 2009, this prospect seems more realistic.

The main question arising out of this phenomenon is: who holds the key to oil price fluctuations? As a corollary, we would wonder as to whether the Governments and academicians have the key to plan for sustainable energy security.

The Strategy

There are two main determinants of change in oil prices. Firstly, new discoveries such as the discovery of the North Sea oilfield in 1986 tend to pull down the oil prices. Short-term supply constraints such as the labour disturbances, or the riots, or wars and threats of conflict tend to pull up the oil prices. In the long term, the real oil reserves tend to determine the annual average of oil price.

Secondly, expectation over the economic growth tends to determine the direction of oil price. Planning for sustainable energy security must take into account nuclear and alternate/renewable sources of energy. This argument is backed up by the 'peak oil' theory proposed by M. King Hubbert—an American geophysicist. The theory states that in a geographical area, oil production initially increases at an increasing rate with the addition of infrastructure and through new discoveries. Beyond this point, oil production increases at a decreasing rate and the fall is due to difficulty in recovering the reserves.

The oil production peaks at a point were about half of world's oil reserves are utilised. The difficulty in recovery is manifested in greater amount of energy spent in recovery. Hitherto, one barrel worth energy enabled the recovery of 50 barrels of oil. Now, only 1 to 5 barrels are recovered depending on the area. For instance, US oil wells consume greater energy than those in the Middle East and Central Asia. Although new techniques like the tar sands, noxal oil, oil shales etc., exist, it is feared that very soon, the net energy recovery would be zero. Studies vary in their estimates regarding the peaking of oil production between 2005 and 2030. Hubbert's peak theory is being extended to the study of natural gas and coal.

Although it is said that natural gas production would peak by 2020 and coal by 2150, use of natural gas in tar sands and the mining of coal with increasing high ash content causes faster depletion of natural gas and coal with low ash content and high energy efficiency. The theory tends to guide the policymakers towards planning for the sourcing of that form of energy whose peak is farther. In the era of lack of consensus regarding the actual timing of the peak, analysis of the options and risks become a major challenge.

In the case of India, a gradual transition is being made towards natural gas *viz.*, the liquefied natural gas (LNG) and compressed natural gas (CNG). Since the peaking of the coal production is likely to be in the mid 21st century according to several estimates, does it mean that a transition should favour coal within the next generation? On grounds of environmental hazards and technical difficulties in running the internal combustion engines, this proposition is least likely to be entertained. Even the 'coal oil' is highly polluting *vis-à-vis* the crude petroleum. (see *http://en.wikipedia.org/wiki/Hubbert_peak_theory, http://www.hubbertpeak.com/* and *http://www.theoildrum.com/)*. Under these circumstances, what should be the appropriate strategy for energy security?

Inferences

The discussion in this section enables us to understand that oil and natural gas can hardly be counted as viable part of India's—perhaps global strategy towards sustainable energy security. Renewable sources *viz.*, the wind and hydropower have better prospects of competing with nuclear power in the next few generations. Only with major breakthrough in the thorium-based technologies, the nuclear power will stand a clear winner.

Despite the discouraging conjecture regarding the India–China competition for oil and natural gas in Africa and Asia, there is a silver lining to the above scenario. China is greatly overpaying for the African and Asian oil which would seriously impede the economic growth. This would also impede China's investment in renewable energy research. Hence, cooperation with India would enable China to obtain oil and natural gas at favourable price. Although India has to pay somewhat higher price for oil and natural gas, at least it will ensure energy security for time to come.

Conclusions and Recommendations

The world is facing various energy risks. They are understood as the systemic risks and the market risks.

The world has to move away from the oil, gas and coal-based to nuclear and renewable-based energy use. Hydropower potential is fast approaching stagnation. Also, the Herfindahl Index shows the extent of source diversification.

India's energy independence is a distant phenomenon which possibly would not be realised within this century. India's strategic disadvantage arising out of geopolitical situation in sourcing for oil has to be overcome by establishing stable and durable economic relationship with reliable partner. Energy efficiency has to be increased and technological advancements in the arena of energy exploration, development and production along with consumer appliances has to be adapted on a grander scale to improve the energy security and to continue the process of reduction in energy intensity.

Indo-Russian defence ties and rapidly growing bilateral trade has to be strengthened through better energy cooperation. There is a possibility to strategise China as a buffer between India and Russia to enhance secure energy imports from Russia to India and foster better business and economic cooperation. China is the biggest trading partner of India with bilateral trade touching US $25 billion. Whereas Indo-Russian bilateral trade will only be $10 billion by the year 2010. By securing the trade routes between India and Russia both overland and through the South China Sea, both India's as well as China's economic security can be ensured on a sustainable basis. To fulfill this objective, Indo-Russian-Chinese naval cooperation has to be fostered through mutual trust and goodwill.

The present Petroleum Minister of India Mr Murli Deora has called upon international community to create strategic business initiatives with across the border investments by oil producer and consumer nations in each other's countries. Deora emphasised the need for helping each other in the exchange and enhancement of technologies in frontier areas like gas hydrates, coal gasification, gas to liquids, coal bed methane, ultra deep exploration, enhanced oil recovery and improved oil recovery techniques. (see *http://pib.nic.in/release/release.asp?relid=46620)*

Appendix A-2.1

Twenty-Eight Year History of Oil, Coal and Natural Gas

Data		*2006*				*2007*
	28 year history	*India*	*Asia & Oceania*	*World*	*Rank*	*India*
Petroleum (Thousand Barrels per Day)						
Total Oil Production		854.19	8,474	84,543	22	881.11
Production of crude oil (including lease condensate), natural gas plant liquids, and other liquids, and refinery processing gain (loss). Negative value indicates refinery processing loss.						
Crude Oil Production		688.61	7,459	73,461	24	697.53
Includes lease condensate.						
Consumption		2,657.80	24,526	84,979	6	F 2,800
Consumption of petroleum products and direct combustion of crude oil.						
Net Exports/Imports(-)		#####	-16,052	—	7	F- 1,919
Net Exports = Total Oil Production – Consumption. Negative numbers are Net Imports.						
Total Oil Exports to U.S.		12	315	13,707	35	29
Total crude oil and petroleum products.						
Refinery Capacity		2,255	22,206	85,345	8	2,256
Crude oil distillation capacity as of January 1. Sources: U.S. data from EIA; Other countries from Oil & Gas Journal.						
Proved Reserves		5.848	36	1,293	20	5.625

As of January 1. Sources: U.S. data from EIA; Other countries from Oil & Gas Journal.

(Billion Barrels)

Data	*27 year history*	*2005 India*	*Asia & Oceania*	*World*	*Rank*	*2006 India*
Natural Gas (Billion Cubic Feet)						
Production		1,056	12,991	101,528	23	1,067
Dry natural gas.						
Consumption		1,269	14,586	103,700	19	1,349
Dry natural gas.						
Net Exports/Imports(-)		-213	-1,535	—	29	NA
Net Exports = Exports - Imports. Negative numbers are Net Imports. Note: Data range begins with the year 1990.						
Proved Reserves		32,525	386,298	#####	24	38,880
As of January 1. Sources: U.S. data from EIA; Other countries from Oil & Gas Journal.						
Coal (Million Short Tons)						
Production	473.172	3,608	6,490	3		497.181
Production of primary coal (includes anthracite, bituminous, lignite, and for Estonia, oil shale).						
Consumption	507.315	3,537	6,483	3		542.787
Consumption of primary coal (includes anthracite, bituminous, lignite, and for Estonia, oil shale) and net imports of metallurgical coke.						
Net Exports/Imports(-)		-1018.8	230	—	6	#####
Net Exports = Exports – Imports. Negative numbers are Net Imports. Includes primary coal and metallurgical coke.						
(Trillion Btu)						
Electricity (Billion Kilowatthours)						
Net Generation		661.64	5,589	17,351	5	NA
Conventional thermal electricity, hydroelectric power, nuclear electric power, and geothermal, solar, wind, and wood and waste electric power generation.						
Net Consumption		488.53	5,075	15,747	7	NA
Net generation+electricty imports-electricity exports-electricity distribution losses.						
Installed Capacity (GWe)		137.578	1,150	3,872	5	NA
One billion watts or one thousand megawatts of electric capacity, as of January 1.						
Total Primary Energy (Quadrillion Btu)						
Production		11.731	115	460	7	NA

Production of petroleum (crude oil and natural gas plant liquids), dry natural gas, and coal, and net generation of hydroelectric, nuclear, and geothermal, solar, wind, and wood and waste electric power.

Consumption	16.205	148	463	5	NA

Consumption of petroleum, dry natural gas, and coal, and net hydroelectric, nuclear, and geothermal, solar, wind, and wood and waste electricity. Also includes net electricty imports.

Energy Intensity4,001	—	—	—	NA

Consumption per dollar of gross domestic product using purchasing power parities.

(Btu per 2000 U.S. Dollars)

Carbon Dioxide Emissions (Million Metric Tons of CO)

Total from Consumption of Fossil Fuels 1,165.72 10,362 28,193 5
NA

Emissions from the consumption of petroleum, natural gas, and coal and the flaring of natural gas.

3 Recent Trends in the Indian Energy Sector

At least 2/3rd of Indians live in rural areas. This is reflected in the fact that bulk of cooking and lighting in rural areas is still based on bioenergy. In urban areas, LPG, kerosene and electricity dominate the scene. India is the net importer of crude oil and several petroleum products. India also is a net exporter of aviation fuel, heavy engine oil, naphtha etc.—an achievement which cannot be overlooked. India's oil refinery capacity has geared up to meet the international market. India's oil refining capacity has reached 18 million tonnes per annum.

India's Electricity Act of 2006 and the Integrated Energy Policy of 2008 have envisaged quality, reliable and optimal supply of energy for all. Universal supply of electricity by the year 2012 and universal utilisation of LPG are the important goals of the Government of India.

This chapter is divided into nine sections. Section II discusses the rural-urban differences in the pattern of energy consumption. Section III discusses the import/export performance of India in petroleum products, and attempts to ascertain the net Government position and the net position of the oil and gas companies in India due to the impact of subsidies and taxes. Section IV discusses the policy and regulatory support provided to the upstream, midstream and downstream oil and natural gas companies in India. The Section goes on to discuss the efforts to increase the oil and natural gas production and the reserve/resource ratio through the MoU mechanism between the Government and oil and natural gas exploring, developing and producing companies. Section V discusses the situation of India's oil and natural gas marketing companies, their investment drive to enhance their national and global presence, risks and challenges faced by them, and Government support provided to them, etc. Section VI discusses the recommendations made by the Committee on Pricing and Taxation of the petroleum product to reduce the subsidy burden. Section VII discusses the demand scenario for the energy items in the household consumption for India up to the year 2031. Section VIII discusses the electricity situation in India in terms of the transmission and the distribution position and the peak time demand/supply position. Lastly, Section IX presents the concluding remarks.

India's Household Energy Use

Bulk of the rural households depend on forms of bioenergy like the cowdung cakes for cooking and kerosene for lighting. About half of the rural households depend on electricity for lighting. Up to 75 per cent of the cooking is done using the bioenergy sources. Only about 9 per cent of the rural households utilise LPG for cooking. Turning to the urban areas, we understand that over 57 per cent of cooking is done using the LPG and over 21 per cent of cooking is done using the kerosene. Ninety-two per cent of the households utilise electricity for lighting.

Petroleum Products

India is the net importer of petroleum products from the mid-1970s. The position had worsened in 1990s and in the first decade of the 21st century. At present, India imports about 70 per cent of its petroleum requirements. Table 3.3 presents India's performance in terms of the actual net import of crude oil and petroleum products.

Between 1990-91 and 2006-07, the imports of the crude oil increased more than four-fold and the import of LPG increased more than eight-fold respectively. The promised production of natural gas from the Krishna Godavari basin did not materialise until 2008. Hence, a sudden fall in the LPG imports in 2000-01 resulted in a sudden bouncing back in the LPG imports in 2002-03.

Although the kerosene imports reduced until 2003-04, we have witnessed a bouncing back of the imports since then. It is commendable that India through the aegis of the Oil India Ltd. (OIL), Bharat Petroleum Corporation Ltd. (BPCL), and Hindustan Petroleum Corporation Ltd. (HPCL) is making all the possible efforts towards self-sufficiency in the byproducts like the furnace oil, lubricants and heavy engine oil, and even in the aviation fuel creating the potential for exports in these areas. In fact, India is making strides in the exports of aviation fuel, naphtha, light high speed diesel etc., as we shall witness in Table 3.4.

Table 3.1

Per cent Distribution of Households by Primary Source of Energy Used for Lighting: All-India

Round	*Year*	*Kerosene*	*Electricity*	*Other/NR*	*No Lighting Arrangement*	*Total*
Rural						
61	July 2004 - June 2005	44.4	54.9	0.1	NA	100
60	January 2004 - June 2004	45.6	53.5	0.6	0.3	100
59	January 2003 - December 2003	46.6	51.7	0.7	1	100
58	July 2002- December 2002	47.4	52	0.5	0.1	100
57	July 2001 - June 2002	47.2	51.9	0.4	0.5	100
56	July 2000 - June 2001	47.8	51.1	0.6	0.5	100
55	July 1999 - June 2000	50.6	48.4	0.5	0.5	100
Urban						
61	July 2004 - June 2005	7.1	92.3	0.1	NA	100
60	January 2004 - June 2004	7	92.2	0.6	0.2	100
59	January 2003 - December 2003	8.3	90.7	0.7	0.3	100
58	July 2002- December 2002	8.3	91.3	0.3	0.1	100
57	July 2001 - June 2002	7.8	91.4	0.3	0.5	100
56	July 2000 - June 2001	9	90.7	0.1	0.2	100
55	July 1999 - June 2000	10.3	89.1	0.3	0.3	100

Note: NR - Not reported.

Source: *TERI Energy Data Directory & Yearbook 2007.*

Government plays a key role in encouraging the consumption of the petroleum products by providing subsidies. Efforts to reduce the subsidy burden of the Government commenced materialising only from the year 2000-01. Subsidy on kerosene increased until 2000-01 and by 2006-07, it is only a quarter of what it was in 1993-94.

Table 3.2

Per cent Distribution of Households by Primary Source of Energy Used for Cooking: All-India

Round	Year	Coke/Coal	Firewood Chips	LPG	Dung Cake	Kerosene	No Cooking Arrangement/ Other/NR	Total
Rural								
61	July 2004 - June 2005	0.8	75	8.6	0.2	NA	1.3	100
60	January 2004 - June 2004	0.7	75.7	9	9.4	0.9	4.3	100
59	January 2003 - December 2003	0.9	74.9	9.1	9.3	1.9	3.9	100
58	July 2002- December 2002	0.9	74.3	8.5	9.9	1.6	4.8	100
57	July 2001 - June 2002	1.2	73.4	8.1	1050	2	4.8	100
56	July 2000 - June 2001	1.3	75.48	7.21	9.51	2.4	4.1	100
55	July 1999 - June 2000	1.5	75.5	5.4	10.6	2.7	4.3	100
Urban								
61	July 2004 - June 2005	2.8	21.7	57.1	1.7	10.2	1.1	100
60	January 2004 - June 2004	2.5	21.5	56.4	1.4	10.4	7.8	100
59	January 2003 - December 2003	3.3	19.98	55.34	1.8	12.99	6.59	100
58	July 2002- December 2002	3.3	21.2	51.2	1.7	14.8	7.8	100
57	July 2001 - June 2002	2.99	23.25	49.8	1.6	15.27	7.09	100
56	July 2000 - June 2001	3.8	21	47.4	1.7	19.4	6.7	100
55	July 1999 - June 2000	4.1	22.3	44.2	2.1	21.7	5.6	100

Source: *TERI Energy Data Directory & Yearbook 2007.*

Table 3.3

Import of Crude Oil and Petroleum Products: 1990/91 to 2006/07

(Thousand tonnes)

Imports	*1990-91*	*1994-95*	*1995-96*	*1996-97*	*1997-98*	*1998-99*	*1999-2000*	*2000-01*	*2001-02*	*2002-03*	*2003-04*	*2004-05*	*2005-06*	*2006-07*
Crude oil	20699	27349	27342	33906	34493	39808	57805	74097	78706	81989	90434	95861	99409	110858
Petroleum, oil, and lubricants														
Light distillates														
Naphtha	—	—	—	—	1874	2407	1917	3165	3308	2784	2371	2214	2473	5474
Liquefied petroleum gas	329	592	678	1035	1226	1722	1587	853	659	1073	1708	2334	2719	2288
Others	5	2	439	455	358	251	—	0	0	—	450	843	882	889
Subtotal	334	594	1117	1490	3458	4380	3504	4018	3967	3857	4529	5391	6074	8651
Middle distillates														
Aviation turbine fuel	27	78	97	150	59	3	1	1	2	2	2	3	2	2
Kerosene	3340	4240	5001	4279	4391	7066	6312	1918	391	698	804	211	881	1425
High-speed diesel	4680	8637	12852	13608	13776	10231	5006	—	31	106	100	814	732	918
Subtotal	8047	12955	17950	18037	18226	17300	11319	1919	424	806	906	1028	6074	2344
Heavy ends														
Furnace oil/low sulphur heavy stock	—	267	1 209	694	634	1696	1378	1728	1425	1256	953	741	789	986
Lubes/others	279	135	58	44	652	396	407	1602	1193	1309	1613	1668	3199	4985
Subtotal	279	402	1267	738	1286	2092	1785	3330	2618	2565	2566	2409	3988	5971
Total imports	8660	13951	20334	20265	22970	23772	16608	9267	7009	7228	8001	8828	11677	16966

Source: *TERI Energy Data Yearbook 2007.*

Table 3.4

Export of Crude Oil and Petroleum Products: 1990 to 2006/07

(Thousand tonnes)

Exports	*1990-91*	*1994-95*	*1995-96*	*1996-97*	*1997-98*	*1998-99*	*1999-2000*	*2000-01*	*2001-02*	*2002-03*	*2003-04*	*2004-05*	*2005-06*
Petroleum, oil and lubricants													
Light distillates													
Naphtha	1520	2249	2461	2184	2048	720	583	2882	2515	2067	2176	2926	8308
Liquefied petroleum gas	—	12	17	22	—	—	—	—	—	—	—	145	86
Mogas	—	—	—	—	—	—	131	1202	2406	2336	2979	2897	3696
Tertiary amyl methyl ether	—	—	—	—	—	—	0	137	87	90	83	0	0
Others	604	609	539	437	—	—	—	—	—	—	210	320	206
Subtotal	2124	2870	3017	2643	2048	720	714	4221	5008	4493	5448	6288	12296
Middle distillates													
High speed diesel/ light diesel oil	82	182	209	216	—	—	—	1597	2890	3178	6181	7286	11645
Aviation turbine fuel	79	170	187	221	—	—	—	160	194	697	1660	2480	3662
Superior kerosene oil	—	—	—	—	—	—	—	—	—	—	—	207	150
Subtotal	161	352	396	437	—	—	—	1757	3084	3875	7841	9973	15457
Heavy ends													
Furnace oil/low sulphur heavy stock	357	29	22	82	333	—	—	508	482	1120	1310	1792	3759
Others	6	3	—	—	—	—	32	1879	1 491	801	21	158	882
Subtotal	363	32	22	82	333	—	32	2387	1973	1921	1331	1950	4641
Total exports	2648	3254	3435	3162	2381	720	746	8365	10065	10289	14620	18211	32394

Source: *TERI Energy Data Directory & Yearbook 2007.*

Subsidy on domestic consumption of LPG increased almost six-fold between 1993-94 and 2000-01 and then dropped down to almost the 1993-1994 level by 2006-07. Subsidy on petroleum products and lubricants as a percentage of the total Central government subsidies dropped down from 34.1 per cent to 5.5 per cent between 1993-94 and 2006-07. This reduction is substantially due to the fact that the total Central government subsidies tripled in the corresponding period while the subsidy on petroleum products reduced rather slowly. Table 3.5 shows the trends in the subsidies on major petroleum products.

Indian Consumer has not benefitted from the petroleum product subsidy. The subsidies are cancelled away by the excise and customs duty imposed on these products as illustrated in Table 3.5.

Highlights of India's Petroleum and Natural Gas Sector during 2009-10[1]

(1) India has total reserves (proved & indicated) of 1,201 million metric tonnes of crude oil and 1,437 billion cubic metres of natural gas as on 1.4.2010 (Table 7).

(2) The total number of exploratory and development wells and metreage drilled in onshore and offshore areas during 2009-10 was 428 and 1,019 thousand metres respectively (Table 8).

(3) Crude oil production during 2009-10 at 33.69 million metric tonnes is 0.55 per cent higher than 33.51 million metric tonnes produced during 2008-09 (Table 11).

(4) Gross production of natural gas in the country at 47.51 billion cubic metres during 2009-10 is 44.63 per cent higher than the production of 32.85 billion cubic metres during 2008-09 (Table 11).

(5) The production of natural gas at 44.94 per cent and 0.08 per cent of the total were highest and lowest in JVC/private and West Bengal respectively during 2009-10 (Table 11).

(6) The flaring of natural gas in 2009-10 at 2.09 per cent of gross production is lower than at 3.29 per cent in 2008-09 (Table 12).

1. *Basic Statistics on India's Petroleum and Natural Gas.*

Table 3.5

Subsidies on Major Petroleum Products: 1993/94 to 2006/07

(Rs million)

Product	*1993-94*	*1994-95*	*1995-96*	*1996-97*	*1997-98*	*1998-99*	*1999-2000*	*2000-01*	*2001-02*	*2002-03*	*2003-04*	*2004-05*	*2005-06*	*2006-07*
Kerosene (domestic use)	37,730	37,400	41,900	65,400	58,200	57,700	81,510	75,220	53,100	30,180	20,180	9,790	9,760	9,670
Liquefied petroleum gas (packed) (domestic use)	12,610	14,100	16,300	22,200	16,600	2,600	44,930	67,240	58,300	36,910	27,830	14,680	15,200	15,680
Total	50,340	51,500	58,200	87,600	74,800	83,700	126,440	142,460	111,400	67,090	48,010	24,470	24,960	25,350
Total Central government subsidies	126,820	118,540	126,660	161,250	185,400	235,930	244,870	268,380	312,100	435,330	443,230	459,570	475,200	534,630
Petroleum, oil, and lubricant product b	65,960	65,600	93,600	186,000	74,800	83,700	177,140	230,910	111,400	53,250	63,510	29,570	26,830	27,850
Total subsidy	193,780	184,140	220,260	260,200	271,240	319,630	422,010	499,330	423,470	487,400	506,74	489,140	502,030	562,480
Petroleum, oil, and lubricant subsidy as percentage of total subsidy	34.22	35.63	42.5	53.56	28.75	26.19	41.98	46.25	26.31	12	14.33	6.43	5.65	5.21

Source: *TERI Energy Data Directory & Yearbook 2007*.

Table 3.6

Realisation of Excise and Customs Duties from Crude Oil and Petroleum Products: 1990/91 to 2005/06

(Rs million)

Product	*1990-91*	*1992-93*	*1993-94*	*1995-96*	*1996-97*	*1997-98*	*1998-99*	*1999-2000*	*2000-01*	*2001-02*	*2002-03*	*2003-04*	*2004-05*	*2005-06*
Excise duties														
Motor spirit	12020	12870	15240	16310	21170	29420	44420	50180	54260	85010	115630	125750	137920	175540
Kerosene oil	2550	2390	3060	2580	6470	2260	2380	2490	7250	6990	13900	17000	12730	2120
Refined diesel oil	6930	7430	11810	11550	19620	20520	31390	67690	81470	110280	105700	134700	144550	217730
Diesel oil	350	340	1070	800	12300	1450	1540	12330	11040	11790	10380	9920	12460	5050
Furnace oil	490	640	1840	1460	2890	5440	3060	6160	9810	7790	9710	8220	9960	17560
Petroleum, oil and lubricant products	4210	4270	5840	7330	8810	18990	22020	27150	33780	28780	34830	29060	37110	3190
Petroleum gases	660	790	2000	2760	3820	3970	3790	3590	12960	15420	24450	25520	24240	46230
Cess on crude oil	27570	21690	26100	28800	26750	27400	26340	32430	36060	27310	45010	51340	52480	50070
Subtotal	54780	50420	66960	71590	101830	109450	134940	202020	246630	293370	359610	401510	431450	517490
Customs duties														
Crude petroleum	31450	40530	34640	40520	57400	48670	32300	62570	77940	48180	68200	74910	97610	71580
Petroleum products	9200	12270	27450	44010	57840	51590	47890	62030	34550	19490	23460	30910	34890	42360
Subtotal	40650	52800	62090	84530	115240	100260	80190	124600	112490	67670	91660	105820	132500	113940
Grand total	95430	103220	129050	156120	217070	209710	215130	326620	359120	361040	451270	507330	563950	631430

Source: *TERI Energy Data Directory Yearbook 2007.*

(7) The refining capacity in the country increased to 184.386 million tonnes per annum (MTPA) as on 1.4.2010 from 177.968 MTPA as on 1.4.2009 (Table 14).

(8) The total refinery crude throughput during 2009-10 at 160.03 million metric tonnes is 0.46 per cent lower than 160.77 million metric tonnes crude processed in 2008-09 and the prorata capacity utilisation in 2009-10 was 89.92 per cent as compared to 107.43 per cent in 2008-09 (Table 14).

(9) The production of petroleum products during 2009-10 was 151.898 million metric tonnes (including 2.244 million metric tonnes of LPG production from natural gas) registering a decrease of 0.51 per cent over last year's production at 152.678 million metric tonnes (including 2.162 million metric tonnes of LPG production from natural gas) (Table 15).

(10) The country exported 50.974 million metric tonnes of petroleum products against the imports of 23.49 million metric tonnes (including 8.828 million metric tonnes of LNG) during 2009-10 (Table 17).

(11) The consumption of petroleum products during 2009-10 were 138.196 million metric tonnes (including sales through private imports) which is 3.60 per cent higher than the sales of 133.400 million metric tonnes during 2008-09 (Table 18).

(12) The total number of retail outlets of public sector oil marketing companies as on 1.4.2010 has gone upto 36,462 from 34,948 on 1.4.2009 (Table 21).

(13) The total number of LPG consumers of public sector oil marketing companies as on 1.4.2010 were 114.952 million against 105.632 million as on 1.4.2009 (Table 25).

(14) The number of persons employed (including contract employees) in petroleum industry as on 1.04.2010 and 1.04.2009 are 129,988 and 138,973 respectively (Table 44).

Oil and gas companies are marred by heavy subsidy burden. On diesel, public distribution system (PDS) kerosene and domestic LPG companies are under-recovering Rs 4.97 and Rs 23.74 per litre and Rs 247.00 per kg respectively. This implies that on a daily basis, the under-recovery is about Rs 235 crore (see Petroleum Planning and Analyses, CII, August 2011).

Oil and gas companies are losing around Rs 86,000 crore annually in spite of the price deregulation. However, the silver lining is that, had the price of petrol and natural gas not been deregulated to the extent achieved so-far, the under-recoveries would have been about 1,30,000 crores.

India's Oil and Natural Gas Sector: Policy and Institutional Setting

India's oil (petroleum) and natural gas sectors are divided into the upstream and downstream segments.

Besides, there is a midstream segment which takes care of oil refining, pipeline and surface transport of oil and natural gas, and the bottling of LPG.

The primary concern for India is the ensuring of the supply security of oil and natural gas, and the sufficiency, and export capability of the refining industry.

Supply security is both physical and strategic. Presently, India supplies only 30 per cent of its crude oil, and 50 per cent of natural gas requirements for domestic use, and imports the rest. Projections reveal that India would import over 90 per cent of its crude oil requirements, and most of the natural gas requirements by 2030. Sincere efforts are being made to maintain the ratio of domestic supply to import percentage. These efforts have already started yielding fruits.

From 5.5 billion barrels, the crude oil reserves of India have increased to about 7.3 billion barrels between 2006 and 2011. Exploration, discovery and development efforts have been taking place domestically on an international scale through various rounds of NELP to a notable extent from 1999.

The scarce foreign exchange gets wasted in the import of oil and natural gas. Hence, efforts are being made to strengthen the refining sector to produce surplus refined products for domestic use, as well as for export purpose. The following subsections discuss India's upstream and downstream policies, institutions and incentive structures and the achievements in the aftermath of the policy initiatives. At the outset, acts and rules governing the sector are mentioned.

Policies Governing the Upstream of India's Petroleum Sector

The New Exploration Licensing Policy (NELP) governs the exploration and licensing of petroleum basins, their development and the production sharing contracts. The Directorate General of Hydrocarbons (DGH) is authorised to formulate, implement, monitor and evaluate various rounds of NELP.

About DGH

The Directorate General of Hydrocarbons (DGH) was established in 1993 under the administrative control of Ministry of Petroleum & Natural Gas through Government of India resolution. Objectives of DGH are to promote sound management of the oil and natural gas resources having a balanced regard for environment, safety, technological and economic aspects of the petroleum activity.

DGH has been entrusted with several responsibilities like implementation of NELP, matters concerning the production sharing contracts for discovered fields and exploration blocks, promotion of investment in exploration and production (E&P) Sector and monitoring of E&P activities including review of reservoir performance of producing fields. In addition, DGH is also engaged in opening up of new unexplored areas for future exploration and development of non-conventional hydrocarbon energy sources like coal bed methane (CBM) as also futuristic hydrocarbon energy resources like gas hydrates and oil shales.

NELP Rounds

NELP was formulated by the Government of India, during 1997-98 to provide a level playing field to both public and private sector companies in exploration and production of hydrocarbons with Directorate General of Hydrocarbons (DGH) as a nodal agency for its implementation. Government of India's commitment to the liberalisation process is reflected in NELP, which has been conceptualised keeping in mind the immediate need for increasing domestic production. To attract more investment in oil exploration and production, NELP has steered steadily towards a healthy spirit of competition between national oil companies and private companies. This has been a landmark event in the growth of the upstream oil sector in India. The foreign and Indian private companies are invited to supplement the efforts of national oil companies in the discovery of hydrocarbons. The development of E&P sector has been significantly boosted through this policy of Government of India, which brought major liberalisation in the sector and opened up E&P for private and foreign investment, where 100 per cent foreign direct investment (FDI) is allowed. Under NELP, which became effective in February 1999, acreages are offered to the participating companies through the process of open competitive bidding. The terms and conditions of this open and transparent policy rank amongst the most attractive in the world.

NELP-I

Under the First round of New Exploration Licensing Policy, Government of India invited bids on 8th January 1999 for 48 blocks for exploration of oil and natural gas. Of these, 12 blocks were deepwater (beyond 400m isobath), 26 shallow offshore and 10 were onland blocks. The bid closing date was 18th August 1999. The production sharing contracts (PSCs) were signed for 24 exploration blocks comprising 7 deepwater, 16 shallow offshore and 1 onland. At present 9 exploration blocks are under operation and 15 blocks have been relinquished.

NELP-II

Under the Second Round of New Exploration Licensing Policy, Government of India invited bids on 15th December 2000 for 25 blocks for exploration of oil and natural gas. Of these, 8 blocks were deepwater (beyond 400m isobath), 8 shallow offshore and 9 were onland blocks. The bid closing date was 31st March 2001. The PSCs were signed for 23 exploration blocks comprising 8 deepwater, 8 shallow offshore and 7 onland. At present 4 exploration blocks are under operation and 19 blocks have been relinquished.

NELP-III

Under the Third Round of New Exploration Licensing Policy, Government of India invited bids on 27th March 2002 for 27 blocks for exploration of oil and natural gas. Of these, 9 blocks were deepwater (beyond 400m isobath), 7 shallow offshore and 11 were onland blocks. The bid closing date was 28th August 2002. The PSCs were signed for 23 exploration blocks comprising 9 deepwater, 6 shallow offshore and 8 onland. The exploration activities are going on in all the 23 awarded blocks.

NELP-IV

Under the Fourth Round of New Exploration Licensing Policy, Government of India invited bids on 8th May 2003 for 24 blocks for exploration of oil and natural gas. Of these, 12 blocks were deepwater (beyond 400m isobath), 1 shallow offshore and 11 were on land blocks. The bid closing date was 30th September 2003. The PSCs were signed for 20 exploration blocks. At present 19 exploration blocks are operating, comprising 9 deepwater and 10 onland. The exploration activities are going on in all the 19 awarded blocks.

NELP-V

Under the Fifth Round of New Exploration Licensing Policy, Government of India invited bids for 20 blocks for exploration of oil and natural gas. Of these, 6 blocks were deepwater (beyond 400m isobath), 2 shallow offshore and 12 were onland blocks. The PSCs were signed for all 20 exploration blocks. The exploration activity is going on in all the 20 awarded blocks.

NELP-VI

A total of 55 blocks were offered during the NELP VI Round for exploration of oil and natural gas in 16 prospective sedimentary basins consisting 25 onland, 6 shallow water and 24 deepwater blocks. One hundred and sixty-five bids from 68 E&P companies (36 foreign and 32 Indian) had participated in the bidding process as consortium/individually. The PSCs were signed for 52 exploration blocks comprising 21 deepwater, 6 shallow water and 25 onland. The exploration activities are going on in all the 52 awarded blocks.

NELP-VII

A total of 57 blocks were offered during the NELP VII Round for exploration of oil and natural gas in 18 prospective sedimentary basins consisting 29 onland, 9 shallow water and 19 deepwater blocks. On 22nd December 2008, contracts were signed for 41 blocks out of which comprised 11 blocks in deepwater, 7 blocks in shallow water and 23 onland blocks.

NELP-VIII

Under the Eighth Round of New Exploration Licensing Policy, Government has signed 31 PSCs on 30th June 2010. There are 8 deepwater blocks, 11 shallow water blocks and 12 onland blocks which are in the states of Assam (2), Gujarat (8), Madhya Pradesh (1) and Manipur (1).

(*Note*: Register on the DGH website to see the names of the blocks, geographical distribution, and strata.)

Sedimentary Basins

India's potential oil and natural gas reserves are assessed in the sedimentary basins. The sedimentary basins of India, onland and offshore up to the 200m isobath, have an area extent of about 1.79 million sq km. So far, 26 basins have been recognised and they have been divided into four categories based on their degree of prospectiveness as presently known. In the deep waters beyond the 200m isobath, the sedimentary area has been estimated to be about 1.35 million sq km. The total, thus, works out to 3.14 million sq km.

Over the last 12 years, there have been significant forward steps in exploring the hydrocarbon potential of the sedimentary basins of India. The unexplored area has come down to 15 per cent which was 50 per cent in 1995-96. To a large extent, credit for this achievement goes to the surveys carried out by DGH in unexplored/poorly explored areas of the country including deepwaters off west coast, east coast and in Andaman sea, and acreages awarded for exploration under NELP. Concerned efforts are continuously being taken to reduce the unexplored area further.

Further four rounds of NELP should be able to completely cover the unexplored areas. However at present, alternate oils *viz.*, the coalbed methane, oil shale, and tar sands are deemed uneconomical in the Indian context given the fact that the prices of alternate oils would shoot up tremendously, and India would not be able to pursue the goal of equity and justice through targetted subsidy.

Besides, the gradual process of increasing the percentage of renewable energy in power and transport is possible only through the marginal taxation on fossil fuel inputs. The Rs 50 cess on a tonne of coal is a beginning towards this direction.

National and International Cooperation in Petroleum R&D

DGH firmly believes that R&D is an essential and integral component of E&P activities. To give a concrete shape to this belief, DGH has initiated cooperation with international and national agencies for undertaking projects in various disciplines related to exploration and production.

National Cooperation

At the national level, DGH has embarked on the following:

1. Upgrade and modify the earth science curriculum to cater to the increasing needs and demands of the E&P industry. In this connection, a draft model curriculum has been framed and the same has been discussed extensively with the faculty and alumni of various reputed universities in India.
2. Three brainstorming sessions were held on 2nd March 2005, 7th January 2006 and 17th October 2008. The Hon'ble Minister (P&NG) chaired all the brainstorming sessions. Five National Working Committees have been formed.
3. Various projects at national level have been initiated through various institutions.

International Cooperation

At the international level, DGH has signed MoU with:

1. Alberta Research Council (ARC), Canada.
2. Energy Resources and Development Unit (ERDU) of Department of Trade and Industry, UK.
3. Norwegian Petroleum Directorate (NPD), Norway.
4. Research Institute of Petroleum Exploration & Development (RIPED), China.

5. University of Aberdeen, UK.
6. Imperial College of London, UK.
7. Department of Industry and Resources, Australia.
8. Japan Oil, Gas and Metals National Corporation (JOGMEC), Japan.
9. The Korea Institute of Geology, Mining and Materials (KIGIAM), Republic of Korea.
10. United States Geological Survey (USGS).
11. United States Department of Energy.

ARC, Canada

MoU with Alberta Research Council, Canada was signed on 26th August 2005 on the following:

1. Estimation of hydrocarbon resources of indian sedimentary basins.
2. National Data Repository.
3. Study of upthrust belts of India.

Work order for all the three projects was placed with ARC. ARC has requested some changes to be incorporated in the work order which was not accepted by DGH. The changes were basically indicating outsourcing of the work to an outside agency to which DGH had reservations. The MoU now automatically stands terminated.

ERDU, DTI (UK)

MoU with Energy Resources and Development Unit of Department of Trade and Industry, UK was signed on 8th September 2005 on the following:

1. Study of frontier basins,
2. Evaluation of development plan involving new technologies.

DTI expressed its inability to spare resources to take up these studies. It was therefore proposed that DTI may outsource the activity pertaining to point no. 1 above or facilitate an agreement between DGH and British Geological Society (BGS) to take up this project. The possibility of signing a separate agreement with the BGS is being explored for the study of frontier basins. The MoU between DGH and BGS has been drafted and is to be approved by MoP&NG/MEA. As regards activity at point no. 2, a seminar was conducted at DTI to understand the latest techniques and technology involved in developing a field and to evaluate the development plan. Two officers from DGH attended the seminar.

NPD, Norway

MoU was signed between DGH and NPD on the 1st of September 2005. DGH has shortlisted two projects. The projects identified are as follows:

1. Assistance in the establishment of National Data Repository (NDR) in India.
2. Better decisions, optimal use of competence and the best possible management of the hydrocarbon resources in India through e-operations.
3. Resource estimation & forecasting.

The activities performed in 2008 were very much in accordance with the plan.

1. Meeting held for two days with representatives from DGH in Delhi to discuss matters related to data management and resource assessment.
2. A group of representatives from DGH visited NPD during the Offshore Northern Seas (ONS) conference in late August 2008. During this meeting, current plans for a new update on the assessment of the petroleum resource base was discussed. It was consented that NPD would support the ongoing work in DGH.
3. Meeting held on February 3-4, 2009 to discuss the establishment of NDR. Also in website restructuring and resource assessment, NPD offered valuable suggestions.

RIPED, China

MoU with RIPED China was signed on January 12, 2006 on the following:

1. Strategy for exploration of sediments covered by volcanic rocks.
2. Improvement in recovery factor in Indian fields by IOR/EOR.

(*Note*: MoU terminated due to no response from RIPED.)

University of Aberdeen, UK

Umbrella MoU was signed by MoP&NG through DGH with the University of Aberdeen, UK on 7th April 2006 on the following:

1. Chemostratigraphy.
2. Geomicrobiology and Micropaleontology studies for identifying chemosynthetic communities.
3. High pressure/high temperature and ultra deepwater drilling.

It was decided that E&P companies would take up the R&D projects directly with the University.

Imperial College of London, UK

Umbrella MoU was signed by MoP&NG through DGH with the Imperial College of London on 7th April 2006 on the following:

1. Controlled source electromagnetic, electromagnetic and magneto-telluric techniques for determining resistivity of subsurface formations.
2. Microseismic monitoring instrumentation in oilfield and passively monitor induced microseismic events.
3. Visualisation techniques to help reduce risk and confidently drill wells with extreme precision and more profitable reservoir management.
4. Thin bed and tight reservoir evaluation.

It was decided that E&P companies would take up the R&D projects directly with the College.

Department of Industry and Resources, Australia

A MoU was signed between DGH of India and Department of Industry and Resources (DoIR) of Western Australia. Stedman Illias, Deputy DG, DoIR and V.K. Sibal, DG, DGH, a technical arm of the Petroleum and Natural Gas Ministry in the upstream sector, signed the MoU.

The MoU aims at enhancing cooperation and understanding on oil and gas related issues and exchange information and pass on good practices in the development of policy on oil and gas exploration, production and field abandonment, exchange of expertise etc. It also identifies broad areas of cooperation. These include reservoir depletion specific to the oil and gas sector, enhanced oil recovery, coal bed methane gas, education and training, joint research and development and any other areas that are mutually agreed upon by both parties. The two sides will seek to approve a programme of cooperation each year or such other jointly decided time period. The MoU is valid for a period of five years and may be further extended with mutual consent.

JOGMEC, Japan

MoU with Japan Oil, Gas and Metals National Corporation (JOGMEC) of Japan was signed on February 16, 2007—progressing satisfactorily.

KIGIAM, Republic of Korea

MoU with the KIGIAM, Republic of Korea was signed on November 25, 2005—lacked response. Stands automatically terminated.

United States Geological Survey

MoU with United States Geological Survey of US was signed on December 16, 2008–progressing satisfactorily.

United States Department of Energy

MoU with United States Department of Energy, US was signed on April 4, 2008 and is progressing satisfactorily.

Additional Policy Guidelines Documents

In addition to the NELP Rounds, and the MoUs to bolster the R&D initiatives, Government of India has been taking measures to bolster the E&P sector. Through various extension and substitution policies. The extension policies are meant to fully cover the unexplored grounds, and to fully assess the commercial viability of the sedimentary basins.

The substitution policies are to make use of the latest developments in R&D in fully utilising the potential of the already dug wells by further improving their recovery factor.

Further, there are policies to facilitate the merger of various rounds of NELP in order to accomplish the unfinished tasks.

Besides, there are policies guiding the determination of the cost of various phases of E&P, classification of hydrocarbon resources, anouncement of discovered fields, submission of application for the expression of interest in projects etc.

- Policy for merger of exploration phases under NELP-III and NELP-IV (shallow and deepwater blocks) production sharing contracts.
- Policy for substitution of additional meterage drilled in deeper wells against total meterage commitment as part of minimum work programme (MWP) in the production sharing contracts.
- Policy for determination of cost of unfinished minimum work programme/committed work programme for exploration blocks under pre-NELP and NELP contracts.
- Policy for extension of exploration phases for exploration and production under coalbed methane contracts.
- Policy for extension in exploration phases in NELP and pre-NELP PSCs and extension thereunder in respect of 23 proposals forwarded by DGH to the Government.
- Guidelines for extension of the exploration phases under the provisions of the extension policy.
- Guidelines for submitting application for assignment of participating interest (PI).
- Guidelines for uniform system of classification of hydrocarbon resources/reserves.
- Guidelines for announcement of new discoveries under production sharing regime.
- Guidelines for extension of exploration phases under NELP and pre-NELP production sharing contracts.
- Guidelines for extension of a phase.

Legal Documents

In addition to the policy documents and guidelines, there are various legal documents governing the development of the oil sector, framing of rules and regulations, ensuring of environmental safety, etc.

Below is the list of the most important legal documents.

- Arbitration and Conciliation Act, 1996.
- Petroleum and Natural Gas Rules, 1959.
- Petroleum Tax Guide, 1999.
- Petroleum and Minerals Pipelines Act, 1962.
- PNG Amendment Rules, 2003.
- Site Restoration Fund Scheme, 1999.
- Oil Industry (Development) Act, 1974.
- Oilfields (Regulations & Development) Act, 1948.
- Petroleum and Natural Gas (Amendment) Rules, 2009.

Laws and Rules Governing India's Petroleum Sector

The oil and natural gas sector is governed by the following acts and rules:

a. The Territorial Waters, Continental Shelf, Exclusive Economic Zone and other Maritime Zones Act (Act No.80 of 1976).

b. The Income-Tax Act (Act No.43 of 1961).

c. The Income-Tax Rules, 1962.

d. The Central Excise Act, 1944 (Act No.1 of 1944).

e. The Central Excise Tariff Act, 1985 (Act No.5 of 1986).

f. The Oil Industry (Development) Act, 1974 (Act No.47 of 1974).

g. The Customs Act, 1962 (Act No.52 of 1962).

h. The Customs Tariff Act, 1975 (Act No. 51 of 1975).

i. The Oilfields (Regulation & Development) Act, 1948 (Act No.53 of 1948).

j. The Petroleum and Natural Gas Rules, 1959.

India's Downstream Petroleum Sector: A Background

The Indian downstream petroleum sector caters to the end user demands. petrol, diesel, kerosene and LPG are the four important end use products.

India maintains price controls on four 'sensitive' petroleum products *viz.*, petrol, diesel, LPG and kerosene in order to insulate consumers from frequent and severe crude price volatilities. The Centrally administered realisable sales prices at which the products are sold are usually much below the 'break-even' prices. The Government does not fully reimburse the under-recoveries of the oil and natural gas marketing companies (OMCs). Government is unwilling to reimburse any under-recovery from this financial year. The reforms aimed at the decontrol of the prices are far less than even to be considered 'half-hearted' as we shall understand in the subsequent subsection.

Government Strategy against Under-Recoveries

Since a structural upward shift in the oil prices began in 2004, the under-recoveries by the oil companies are expanding significantly. Even compared to the 'trade parity' or the 'refinery gate cost', the under-recoveries are significantly high. In 2008-09, the under-recoveries amounted to over US $25 billion or over Rs 1,15,000 crore.

Until the year 2008-09, Government financed up to 80 per cent of the under-recoveries through the 'off company' oil bonds and through debt securities. Certain proportion of the debt securities are used to raise liquid cash and certain proportion are used as collateral to raise loans. Thus, the Government deficit is significantly augmented by the Government's reimbursement of the under-recoveries of the OMCs. In fact, reimbursement of the under-recoveries amounted to over 50 per cent of the Government deficit in 2008-09. This is a stark warning in the wake of India marching towards becoming third largest consumer of petroleum products with a rapid annual increase in consumption of 5.2 per cent.

Figures (3.1-3.3) show the crude price index, the extent of under-recoveries, and the comparison of magnitudes of under-recoveries, taxes and oil bonds.

Figure 3.1

Indian Crude, Product Price Index

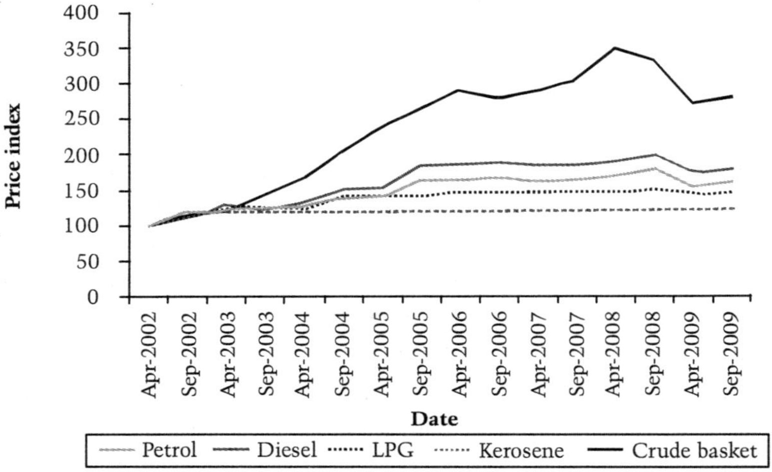

Source: See Kieran Clarke; IEA 2010.

Figure 3.2

Under-Recoveries, 2004-2009

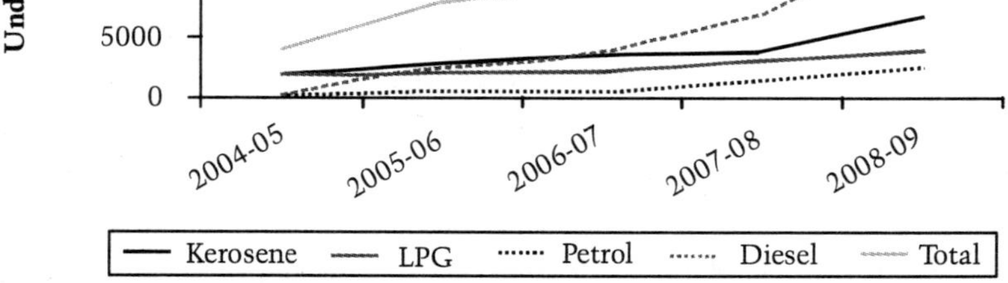

Source: Same as Figure 3.1.

Figure 3.3

Comparing Magnitudes: Oil Sector Taxes, Under-Recoveries and Oil Bonds

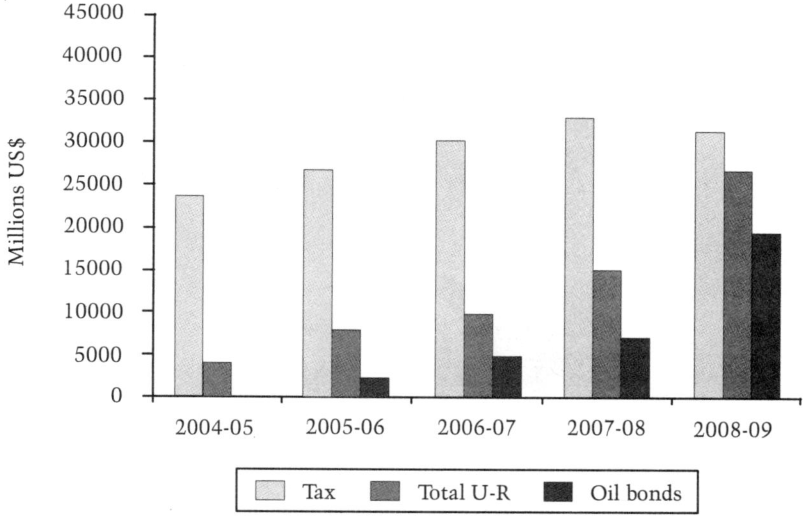

Source: Same as Figure 3.1.

India's Downstream Petroleum Sector Policy

There are two key objectives behind the policy:

(a) Ensuring the sustained supply of refined products at affordable prices and meet the growing demand; and

(b) Establishing India as a major global exporter of refined products.

Between 2007 and 2012, Indian OMCs are investing strongly in the upgradation and expansion of oil refineries. Up to 6,00,000 barrels of greenfield and 3,00,000 barrels of brownfield daily refining capacity is being added which is mostly to meet the export requirements. Private sector is a major contributor to India's refining capacity thanks to its autonomous robust investment capabilities. Reliance Group's Jamnagar-II and Essar Group's upcoming Vadnar-II alone contribute to 1.2 million tonnes or a fifth of India's refining capability. These refineries supply to OMCs on a wholesale basis at competitive prices. India will surpass Singapore and become Asia's largest refined product exporter by 2012 and remain so during the foreseeable future. Product management capabilities, product flows and the supply chain will get strengthened.

Figure 3.4 shows the refining market share.

Figure 3.4

Refining Market Share

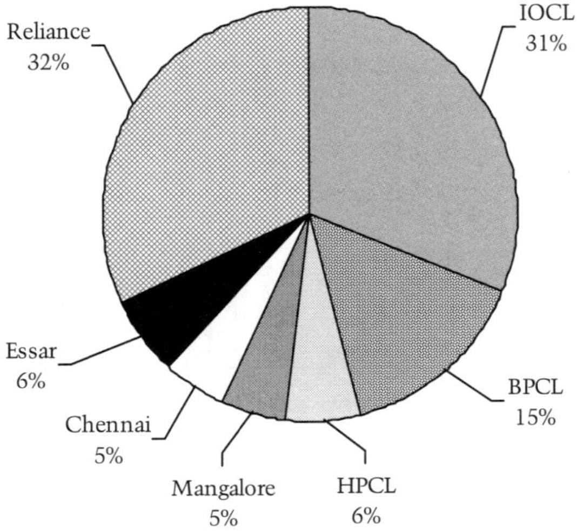

Source: Same as Figure 3.1.

High-end consumer products and clean transport will become a daily reality in India and in the Indian Ocean Rim (IOR) region. Besides, the IOR will enjoy significant increase in its energy security.

Developed countries will import the refined products to meet the incremental demand.

India's Petroleum Pricing Mechanism

Prior to 2002, India had the administrative price mechanism determined on a 'cost of capital plus' formula whose prime motive is to subsidise the consumers. However, this was replaced by the 'import parity price' according to which, the price has to change along with the change in the price of the international crude basket. However, the lifeline requirements of India's low income groups would be jeopardised if the prices of kerosene and LPG are liberalised. Similarly, diesel is consumed by large section of India's motoring population, and agriculturists and industrialists for power generation on account of severe shortage of electricity. Diesel occupies 1/3rd of the crude basket. To protect the interests of low and middle income population, the administered pricing was allowed to return although unannounced. From 2006, trade parity pricing began dictating the price of India's petroleum products based on the difference between the import price of the crude and the export price of the refined products.

Under-recoveries have escalated sharply in recent years. From less than US $10 billion in 2004-05, the figure rose to about US $30 billion in 2008-09.

Under-recoveries are bound to occur if the crude price increases above US $55. From 2007, the price is almost always above US $55 often crossing US $90.

Under-recoveries led to the issuance of the oil bonds. Tradable oil bonds have a maturity period of 5-7 years and some even 20 years. The fiscal deficit is increasing on account of the maturity of the tradable oil bonds beginning from 2010.

Non-tradable oil bonds are used as the collateral banks to lend to the OMCs. However, they do not have the status of statutory liquidity ratio (SLR). Hence, failure to trade them in the secondary market desists the banks from accepting them as collateral. Besides, there is a glut in the bond market. Hence, despite the falling interest rates in 2009, bond prices experienced a sharp fall. The losses to OMCs on oil bonds prompted the Government to mandate RBI to absorb the bonds and also to swap bonds for Forex to facilitate the import of technology and inputs from abroad. The robust technological investment is thus characterised by increased Government debt.

There is thus an emerging issue of 'vertical fiscal imbalance' within India's system of petroleum market regulation. On the one hand, the GoI is systematically cutting taxation revenues while rapidly increasing outlays (in the form of oil bonds) to deal with inherent petroleum pricing issues. On the other hand, state governments continue to reap tax revenues from the petroleum sector while offering no financial support for the crippling system of effective subsidies.

The Cost of 'Effective' Subsidies

Under-recoveries to OMCs is due to the effect of significantly lower retail product prices than crude input prices which is the result of the subtraction of the subsidy from the 'trade parity cost'. It represents the realisable sale value. As per the 2008-09 prices, any price of crude above US $55-60 results in the under-recoveries. Under-recoveries trippled from US $10 billion to US $30 billion between 2004-05 and 2008-09. Between April and December 2008 alone, the top three OMCs in India lost between 25 per cent and 43 per cent of their net worth.

Hence, the Committee on the Petroleum Product Pricing recommended the elimination of under-recoveries, bolstering of liquidity and the infusion of corporate flexibility among the OMCS to make them internally competitive. Oil bonds and a discounted supply of the crude from the upstream PSUs were mandated by the Government. However, the upstream PSUs need to be aggressive in the acquisition of the oil and natural gas assets in India and abroad to improve the crude availability. The discounted supply of oil and natural gas would prove counter-productive in this respect.

Infusion of Liquidity: A Short-Term Measure

To tide over the immediate liquidity crisis and to prevent the erosion of the net worth of the OMCs, Government began issuing the off budget oil bonds. OMCs are expected to convert these bonds into liquid cash. However, the general glut of bonds in the markets has caused saturation. Hence, there is a continuous fall in the bond value. Besides, the bonds are not accepted by the RBI as SLR. This prevents OMCs from borrowing from the RBI at preferential interest rates.

Besides the US $20 billion of oil bonds issued in 2008-09, close to US $10 billion worth fertiliser bonds were issued by the Government reflecting Union Government's inability to subsidise the farmers on fertilisers. Thus, the glut in the bond market results in a continuous fall in the bond prices and the OMCs are not able to fully recuperate the erosion in their net worth. The unwillingness of the RBI to accept the oil bonds as the SLR leads to a competition in the bond market. Even the Government bonds experience loss in value and the ability of the Government to borrow gets eroded. Indian Government's debt is 82 per cent of GDP which includes those of the state governments. Close to 24 per cent of the general government revenue is spent on the interest payments. To worsen the situation, the tax realisation of the Union government on the petroleum products is hard hit.

Rationalising the Tax Structure

Government has substantially brought down the taxes on the petroleum products. Ad-valorem tax on the products is abolished. Only a flat rate of Rs 13.35 on petrol and diesel exists. There is no tax on LPG and kerosene unlike the 8 per cent and 16 per cent in the past. In contrast, the state governments are reaping enormous dividends in the form of the excise and sales taxes. Thus, the Central government's revenues are substantially eroded. This, coupled with substantial stimulus spending is causing enormous fiscal burden on the Government. The medium-term creditworthiness of the Government of India is seriously questionable according to the Standard and Poor (March, 2009). However, the analytical bias of the Standard and Poor is mitigated by the fact that India has successfully weathered the 2008-2011 economic downturn in the industrialised countries and has brought down the fiscal deficit and the current account deficit substantially. The recent 2010-11 budgetary commitment to phase out the subsidy on the petroleum products has an environmental angle. Government has realised that continuous subsidisation of the petroleum products hinders the financing of the clean energy projects. Hence, rather than subsidising the CO_2 enhancing products including the petroleum and coal, a clean energy cess is being imposed to finance the R&D on clean energy—especially the solar and biomass energy. Besides, the attempts to bring the petroleum product subsidy onto the budget is a marked shift from the earlier policy of indirect subsidisation.

India's downstream petroleum sector is thus geared up for market-based competition and the consumers are thus geared up to face the regime without subsidy in a gradual manner whose pace needs to be increased substantially.

Regulatory Structures and Institutions

The Petroleum & Natural Gas Regulatory Board (PNGRB) was established in 2007 as the downstream sector regulator, tasked with regulating the refining, processing, storage, transportation, distribution, marketing and sales of petroleum products and natural gas. It does not, authorise the construction of the refinery infrastructure, which in turn is controlled by the MPNG. The board also has no role in market pricing, or in the pricing policy. The key practical function of the board relates to:

a. Arbitration for dispute settlement within the downstream sector; and
b. Its powers to release tenders, and to authorise the laying, building, operation and the expansion of the natural gas distribution networks.

PNGRB prevents monopolistic practices, registers, marketing and retailing entities, monitors and prevents adulteration, authorises the operators to lay product pipelines, and methodically determines the private or the public nature of these pipelines. The board also regulates access to pipelines and pipeline transportation rates. The board also monitors prices through the downstream value-chain, and the adherence to maximum prices set by the GoI. The board determines and enforces technical standards and specifications relating to downstream activities. Although the legal transparency, transparent regulatory oversight and arbitration are introduced, the 'toothless' nature of the board is revealed in its lack of authority over the product and refinery pricing.

Role of Indian Oil Companies in Meeting the Demand for Refined Products in the Domestic and International Markets

Government *viz.*, the public sector companies, as well as, the private companies play a key role in meeting the demand for refined products in India.

In so far as the downstream sector is concerned, the same Government companies running and expanding their refinery assets are involved in marketing the end products.

Private sector is, by and large, confined to building and operating the refineries, besides undertaking exploration, development and production from the oil and natural gas fields complimenting the efforts of the corporate giants *viz.*, the Oil and Natural Gas Corporation (ONGC) and Oil India Ltd. (OIL).

OMCs have obtained mixed results in their business. OMCs suffered considerable financial losses due to government imposed caps on the retail prices. For each of the OMCs, operating cashflows in 2008-09 were significantly negative. Government support was required to keep the OMCs solvent and marginally profitable. Commercial viability of OMCs was undermined by the administered pricing mechanism and the accompanying 'heavy-handed market regulations'. The resulting dent in the revenue generation capability reduced the investment potential of the OMCS substantially.

OMC sector is defined by heavy-handed regulation and poor core commercial performance. Yet the sector as a whole is characterised by robust investment, capacity addition and growth. This scenario more than satisfies India's oil and natural gas requirements in the medium term *viz.*, until the year 2016. This enhanced investing capability is due to the unique relationship of OMCs with the Government discussed above. Besides, lenders are satisfied with the AAA credit rating bestowed upon the three leading OMCs due to explicit Government guarantee of solvency and profitability in spite of heavy indebtedness and poor balance sheet position.

The greenfield OMC capacity addition is to the extent of 6,00,000 bbl/d apart from the brownfield capacity addition and upgrades and the upcoming ambitious private sector capacity addition. Besides, the Jamnagar greenfield refinery has contributed to India's surplus refining capacity enabling the exports to neighbouring countries as well as to the Europe and to the USA depending on the need.

Government bails out the OMCs in return for reliable market supply and investment expansion on a sustained basis. The first decade of 21st century thus manifested rapid and over-investment from the perspective of domestic demand.

The current system of inter-relations between the Government, OMCs and product markets is highly successful from a policy outcomes perspective. assisted by OMCs, the Government is able to pursue its official policy of providing affordable energy for India's developmental needs and to its significant poor population. How long can the Indian government continue to spend hundreds of billions of Rupees on perpetuating the solvency and investing potency of OMCs under current product pricing regime?

India is emerging as the exporter of the refined products. This is likely to significantly impact on the regional product markets. The depth of product markets expands in the Middle East and in Southeast Asia. Indian refineries would have to compete with the established Asian refinery hubs such as Japan, South Korea and Singapore. The United States Gulf Coast and Northwestern Europe would receive refined products from India because their 'brownfield' refineries can not expand on economic grounds The energy security of Iran, Saudi Arabia, Indonesia, Vietnam and Australia would benefit significantly due to exports from India.

India's Export Opportunities in the Asia-Pacific and in the Middle East

Commercial and policy initiative to establish India as a major refined product exporter is derived from a strong medium- and long-term forecasted growth in demand in the Asia-Pacific and Middle East regions. Besides, Asia is emerging as the leading consumer of the liquid fuels. According to the 2008 *World Energy Outlook*, about 85 per cent of incremental demand for oil until 2030 will occur in Asia-Pacific and the Middle East. Until the year 2015, Asia's average growth in demand for liquid fuels is expected to be 4 per cent per year excluding the Northeast Asia. China's growth in demand is expected to be 4.5 per cent per year.

Figures 3.5 and 3.6 show the growth in the demand for refined products in India, and in Asia.

Figure 3.5

Indian Product Demand

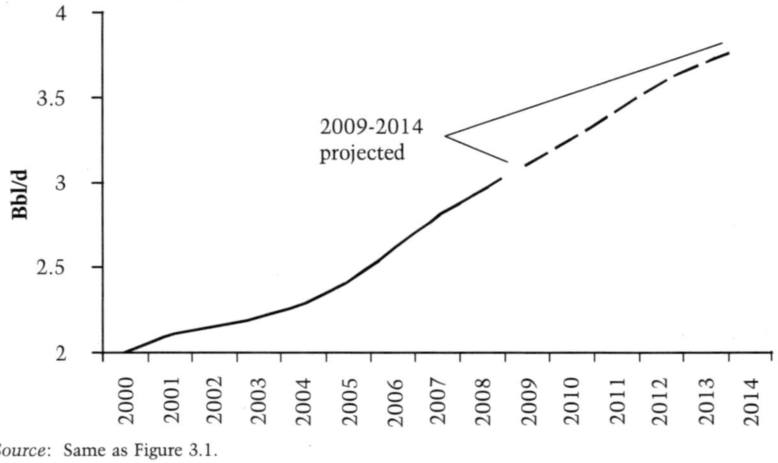

Source: Same as Figure 3.1.

Figure 3.6

Strong Asian Product Demand

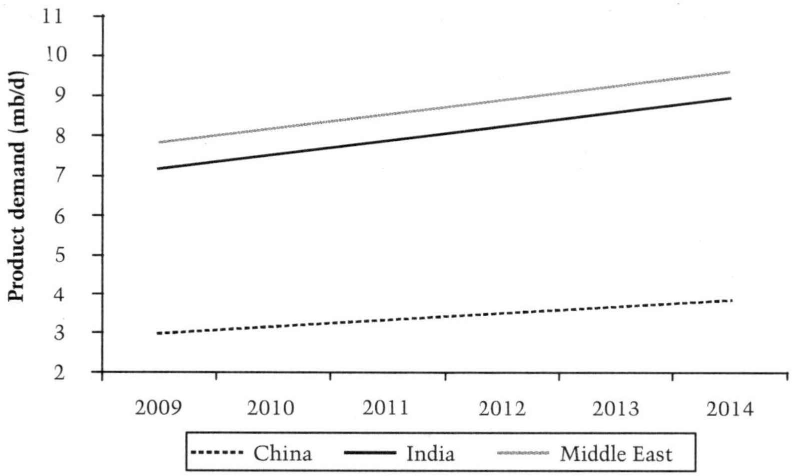

Source: Same as Figure 3.1.

India has good opportunity to export the Euro-V grade fuel to the USA, Europe, Japan and to the Australia. Reliance's Jamnagar-II and Essar's greenfield refineries have geared up to meet this challenge. No greenfield refinery is coming up in the foreseeable future in the North Atlantic region. No refinery in the North Atlantic Region, Japan and in Australia can match the size, complexity and modernity of the Indian greenfield refineries. Close to two million bbl/d of refining capacity is under threat of closure in Europe due to new, cheap, large scale and complex refineries being established in the developing countries, particularly in Asia according to the *World Energy Outlook 2009*.

Medium-Term Refinery Capacity Addition in India

India's refining capacity addition, sophistication, orientation and global importance has grown almost at a revolutionary pace and proportion in the Eleventh Five Year Plan. The expected growth is about 40 per cent from 3.54 to about 5 million bbl/d. OMCs, the private refineries and the downstream companies, will add 6,00,000, 5,00,000 and 3,50,000 bbl/d respectively. This is almost 2.5 times the accumulated capacity addition in Malaysia. Between Jamnagar-II and Vadinar-II *viz.*, between 2008 and 2012, about 1.2 million bbl/d of new refining capacity will be added by the private players. This is highly significant, making them the major players *vis-à-vis* the public sector OMCs. This equals the entire refining capacity of Singapore, a refining powerhouse in Southeast Asia.

Rapid Refining Capacity Addition in the Face of a Suboptimal National Product Pricing Regime

As noted above, private players play a key role because, they are not affected by the under-pricing of the products. By 2012, two million bbl/d of capacity would be operated by the private players. Two key factors could explain this phenomenon.

i. The Role of OMCs

OMCs buy the products from the private refineries at the market based refinery gate prices which assure profits to the private refineries. Besides, the private sector refineries are therefore able to market the industrial fuels, lubes, jet fuel, etc., at market rates without restriction while the public sector OMCs continue to suffer from under-pricing of the four key products *viz.*, petrol, diesel, lpg and kerosene.

ii. India as an Export Base

Private players have considerable incentive to establish export oriented refineries. They are not entangled by the regulatory complexities and the resulting losses unlike the public sector OMCs. India is in proximity to key crude producing regions in the Middle East and the key centres of demand growth *viz.*, the South-East and the East Asia. Besides, the cost of inputs is significantly lower compared to the developed countries and access to the financial markets and the high-tech capital is relatively abundant.

Mukesh Ambani, Chairman of RIL, famously boasted that Jamnagar-I was built 'with double the men, at half the cost and in half the time' typically expected for large-scale refinery projects. Jamnagar-II was completed in 36 months from concept to commissioning—a global benchmark for a refinery of its scale and complexity.

IOCL's 300,000 bbl/d greenfield refinery project, employed a temporary construction workforce of over 1,00,000. Such a large-scale mobilisation would be almost impossible in almost all the nations with the exception of India, China and Indonesia.

After the approval of the construction of Jamnagar-II, Mr Jeet Bindra, Chevron Texaco's Head of Global Refining, admitted that establishing refineries in developed countries was quickly becoming unviable when matched against huge Asian complexes such as Jamnagar. Hence, Chevron took a small 5 per cent stake in Jamnagar-II considering the imports into USA as better option than establishing new refining capacity in USA.

First Mover Advantage for Indian Companies

Indian companies are enjoying the 'first mover' advantage in the refining sector. Besides the public sector OMCs having experience in the brown, as well as in the greenfield refining operations, and having mature domestic trade and distribution networks, the major private player *viz.*, the RIL has built up the same in the international market alongside the construction of the Jamnagar-I and Jamnagar-II refineries.

This has thwarted the international majors from acquiring the refining stakes in India's greenfield refineries. Total S.A. pulled out of the HPCL's 3,00,000 bbl/d refining complex at Visakhapatnam in October 2009 citing falling demand for petroleum products. Exon Mobile pulled out of the 1,80,000 bbl/d Bhatinda project in 2,000 and BP pulled out within six months, despite agreeing for a 50/50 stake in 2005 in the same project. The international majors only have stake in the industrial fuels and lubes markets in India. Shell Oil alone has the permit to establish the retail outlets in India.

International majors, at this stage cannot enter the Indian market and the refining sector due to the incumbents having highly embedded, well established commercial associations, and supply chain linkages. International majors will opt for investment in a location if deeper market penetration, and the introduction of the global brands is permitted by the Indian market conditions.

The product pricing regime in India, and the nature of capital market access to India's OMCs do not at this stage permit them to enter the Indian market. There might be an opportunity after 2015, when the demand for petroleum products pick up on a scale that lets the international majors to enter into the Indian product and refining market.

Besides, there would be an opportunity for smaller domestic players, like the Essar group to expand. RIL's Jamnagar-II has effectively crowded out new investments until 2015.

Figures 3.7 and 3.8 show the India's domestic demand and export capability until the year 2015, along with the export capability of the Asia-Pacific countries.

Figure 3.7

Indian Refining Capacity versus Domestic Demand, 2009-2012

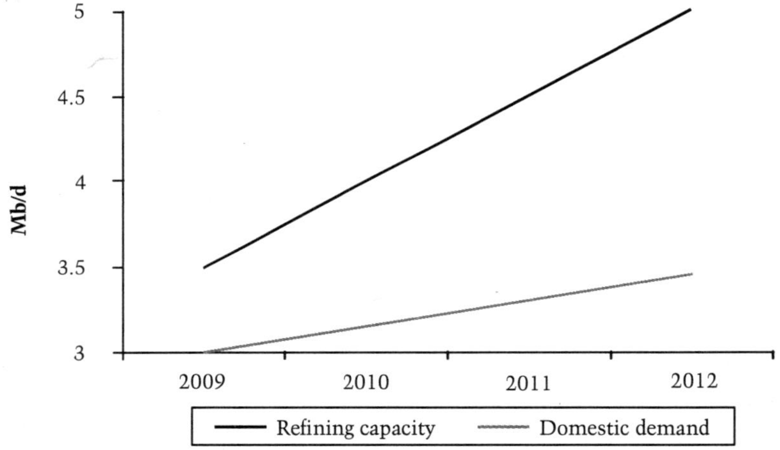

Source: Same as Figure 3.1.

Figure 3.8

Forecast Asia-Pacific Refined Product Exports, 2012

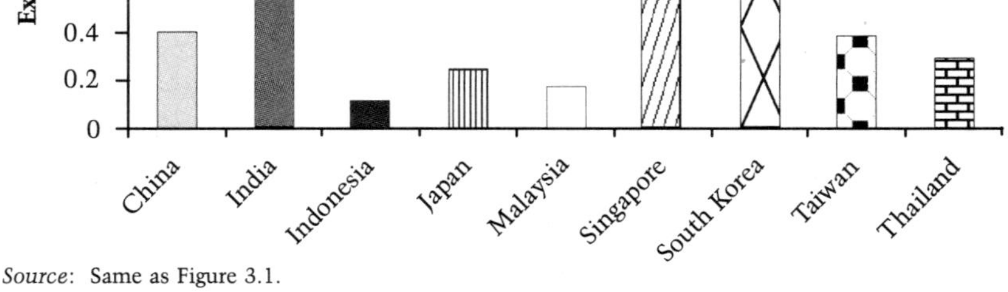

Source: Same as Figure 3.1.

Figure 3.9

Indian Refining Capacity versus Domestic Demand, 2012-2015

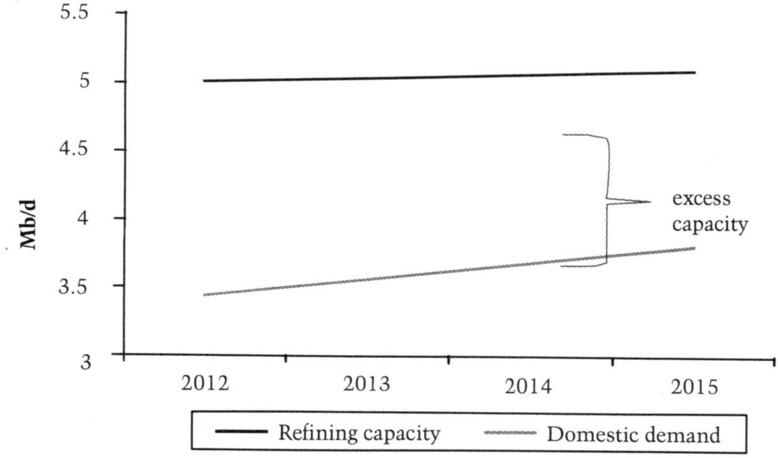

Source: Same as Figure 3.1.

Case Studies on India's Downstream OMCs

IOCL

IOCL is forging ahead with a massive downstream investment and expansion programme to cater to the growing demand in India. IOCL enjoys a relatively debt-free balance sheet. Besides, the Government provides continuous backing. IOCL's investment expenditure was US $2.24 billion, including approximately US $1.2 billion on capacity expansion of the refineries, yield and quality improvements and marketing operations.

This compares with the profit after tax of US $900 million clearly vindicating the Government backing. The medium-term investment strategy is resulting in the investment expenditure of US $9.24 billion, of which, about 60 per cent, or US $5.1 billion is spent on the refinery capacity expansion.

Expansion projects include the construction of a large, state-of-the-art greenfield 300,000 bbl/d refinery at Paradip in Eastern India, with US $4 billion. The project aims to bridge the deficit of light middle distillates in Eastern India, and is to be commissioned by late 2012.

IOCL's Panipat refinery in Haryana is being expanded by 60,000 bbl/d, with a cost of US $300 million to cater to the growing needs of the Northwest India. A long-distance pipeline linking the Panipat refinery with Mundra port in Gujarat is part of this project.

IOCL is investing heavily in petroleum quality upgrade projects at its Gujarat, West Bengal, Haryana, Bihar and Assam refineries to ensure Euro-IV grade fuels by 2011 at the cost of US $2.1 billion.

BPCL

Despite a high level of debt burden, growing interest burden, weak profits and negative net cash flows, BPCL is forging ahead with fair amount of investment plans with the backing of the Government. The regular extensions of oil bonds amounting to about US $5 billion help smoothen the cash flows, bolster the balance sheet, and undertake short-term debt servicing to ensure investor, and creditor confidence.

The greenfield 1,30,000 bbl/d Bina refinery as a joint venture with Oman Oil, the crude receipt and dispatch facilities on the coast at Jamnagar, and a 935 km cross-country crude pipeline from Jamnagar to Bina are the projects under construction.

BPCL has invested over US$ 1.2 billion in the Bina project. BPCL is undertaking a US $1 billion upgrade of its Kochi refinery, to be commissioned in 2011 adding 50,000 Bbl/d, providing Euro-IV products. BPCL is investing to upgrade publicly-owned Kochi port to allow the mooring of very large crude carriers and to establish an undersea pipeline link to the refinery. The total investment is about US $5 billion.

HPCL

In spite of heavy losses, the Government backing ensures the expansionary investment plans of HPCL. The Guru Gobind refinery at Bhatinda in Punjab province—with an initial capacity of almost 200,000 bbl/d is being constructed with the cost of about US $3 billion. Mittal Energy is the co-owner of the refinery. A 900 km pipeline is being constructed from Mundra to the project site. Significant capacity addition is taking place pertaining to the jet fuels, middle distillates, and other petroleum products.

India's Role as the Exporter of Refined Products

Out of the total production of 4.96 bbl/d, only 3.44 million bbl/d would be consumed in India. India would export 1.5 or at least, 1.4 million bbl/d. India thus becomes the fourth largest producer and Asia's largest exporter. Industrial gasoline, alkylate, ultra-low sulphur diesel and pet-coke would be produced in small quantities.

The emergence of this large-scale and complex refinery capacity in India will greatly increase the resilience of regional supply chains for the high-end industrial and auto fuels. This will facilitate the cycle and season nutral exports, in turn contributing to the deepening of the trade flows from Africa to Australia, via the Middle East, India and Southeast Asia. The flow would be very high in diesel, jet fuel, gasoline, complex high-end products leading to the strengthening of the supply chains.

India has the opportunity to export one million bbl/d to the Asia-Pacific and a fair amount to the USA and Europe contributing to the trade flows and the supply chain flexibility. In the context of Southeast Asia, this a significant jump of 20 per cent in the imports.

On the whole, the supply chains are going to become robust and the trade routes more resilient serving both mature economies, and the small emerging economies.

The new open economies, thus enjoy comparative advantage in establishing and operating the refineries. Extremely slow growing mature economies, such as the Western Europe, Australia, the USA, and Canada etc., cannot afford to make costly, lengthy, labour intensive and regulation intensive refinery investments and upgrades. The Indian Ocean region is filling up the refinery investment gap. In the mature economies, no green-field refinery was built since the mid-1980s.

Risks and Challenges to India's Refining Industry

a. Possible Regional Structural Excess Capacity

The major users of oil are experiencing a decline in the demand for refined products due to increased energy efficiency and the increased use of the alternatives like the fuel cells.

This apart, China added a refining capacity of over 1.5 bbl/d in the past 2-3 years. With the excess capacity in smaller refineries, such as the Thailand and Taiwan, excess capacity has almost tripled from 1.4 million bbl/d in 2008 to about four million bbl/d in the Asia-Pacific region. This will grow faster in this region by the year 2015. The Middle Eastern deficit in the transportation fuels will be more than wiped out after 2014 when over 2 million bbl/d of excess capacity would be commissioned. If China's demand growth is more than expected, the refining industry in the region can mitigate the risk of excess capacity, and the resulting low margins, lower throughputs, commercial hardships and the possible downsising.

In the case of India, larger refineries are capable of handling cheaper and heavier crude at lower input and refining costs. Besides, India has the best facilities to refine the high-end products that can be sold globally at a premium. This twin advantage, coupled with lower labour and capital input costs can make India's refineries offer greater competition to the existing older brownfield refineries of Japan, South Korea, etc.

b. Falling Global Demand

The excess capacity is leading to uncertain investment climate, unstale revenue scenario and cyclical excess capacity. This will hinder the resource availability for investment. After 2012, India's economy would grow in a robust manner and the world will also experience strong economic growth sometime thereafter. In the present scenario of lack of refinery investment in India after 2012, India would only export about 1.2 million bbl/d in 2016 as compared to 1.4 bbl/d by Singapore.

Public sector OMCs are likely to face greater resource crunch because the Government would not be willing to support the investment operations in the long run. Besides, resistance to oil price liberalisation is expected to continue for a long period of time.

c. Transformation of India (Mumbai) into a Liquid Fuel Service Hub

The long-run commercial viability of the liquid fuel industry in India hinges upon the establishment and the operation of the shipping, handling, storage and port services along with the allied commerce, financial, trading, management and regulatory services.

India has to emulate the success of Singapore in this respect, and provide such services for the Indian Ocean Rim as a whole.

The Committee on Pricing and Taxation of Petroleum Products

The 'Committee on Pricing and Taxation of Petroleum Products' under the chairmanship of Dr C. Rangarajan identified that without Government intervention, the subsidy burden borne by the OMCs and the upstream companies like the ONGC, OIL and GAIL on the sale of kerosene and LPG cannot be lifted. Consequently, companies lose profits, and the capability to expand the upstream oil and natural gas fields, refineries and downstream marketing. As a result, Government loses tax income and dividends.

The heavy subsidy burden eroded the profits of the OMCS and had negated the profits in spite of huge assistance from the upstream companies which also resulted in the erosion of their own reserves by over Rs 10,000 crore, or $2 billion in the year 2004. Hence, the Committee was constituted to help transition the OMCS towards the market-based pricing of petroleum products and LPG.

The following principles informed the decisions of the Committee.

[1] Pricing and taxation of petroleum products should be rationalised to transmit the right price signals so as to minimise if not eliminate distortions and inefficiencies that result in misallocation of resources.

[2] Prices of petroleum products should, as far as possible, be aligned with international prices.

[3] Across the board subsidies result in inefficiencies and place an undue burden on an already strained fiscal situation. Subsidies should be minimal, targetted and restrained by a monetary ceiling.

[4] To the extent the Government decides to extend subsidies, the burden should be borne entirely and transparently in the Union Budget. The oil marketing companies should be freed from the burden of subsidy.

[5] Custom tariffs on crude and products should be rationalised so as to moderate the effective rate of protection to a level that will offset the disadvantages suffered by the domestic producers without, at the same time, allowing them any undue cushion. Excise tariffs should be restructured to protect the consumers from excessive volatility in prices.

[6] An appropriate pricing regime which promotes efficiency needs to be evolved in relation to petrol and diesel on the one hand and domestic LPG and PDS kerosene on the other. However, it is the latter which is arguably more intractable because of the heavily subsidised prices to consumers. The issues of adjusting prices and targetting them appropriately become urgent in this context.

The committee made the following recommendations to reduce the subsidy burden while at the same time mitigate the burden of price rise on the consumers on account of transition to market-based mechanism.

a. Freight equalisation should be removed so that the cost differentials on account of difference in the freight rate across distances can be captured.

b. Trade parity price should be adopted instead of the import parity price. The weightage given in the later form of pricing is 80 per cent of the cost of import of crude oil, and 20 per cent of the cost of export of finished products. This is analogous to the inclusion of asset replacement cost instead of depreciation cost. Trade parity pricing reflects the changing price of international crude basket.

c. Ad-valorem duties should be replaced by specific duties and cess so that consumers will not be unduely burdened. Octroi, or the entry tax should be removed, and an equivalent surcharge on sales tax should be imposed so that the local governments can be compensated by the state governments.

d. Customs duty should be reduced and an equivalent amount of excise duty should be imposed by the Union government so that consumer would not be overburdened by the tax.

Thus, any price increase on account of the trade parity pricing would be mitigated. It is found that 38 per cent of PDS kerosene is diverted to non-PDS purpose. Preventing this diversion can save the subsidy bill by a third. Lpg being mostly consumed by the above poverty line (APL) population, increase in the cost of cylinder by 40 per cent will eliminate the subsidy bill by 40 per cent. The efforts towards eliminating the LPG subsidy is marginal.

While the price of LPG cylinder increased by 73 per cent in the past six years, to Rs 405, the cost of the cylinder also increased by 75 per cent to Rs 700. The percentage of LPG consumers almost doubled, thus more than doubling the LPG subsidy. Government hence has to take more courageous steps to eliminate the LPG subsidy, or at the most target the subsidy to the really BPL families.

Household Energy Demand Scenario in India

The energy items demand forecast until the year 2031 under the 8 and 9 per cent GDP growth rate scenario suggest that there is a huge unmet demand for electricity and LPG. Hence, electricity and LPG consumption are expected to grow four- and eight-fold respectively. Conversely, the demand for firewood and chips and the kerosene are expected to grow by about 33 per cent. This suggests that as the economy grows and as the standard of living improves, significant proportion of population demand comforts and hence are inclined towards the electricity and LPG. Tables 3.7 and 3.8 illustrates this phenomenon.

Table 3.7

Demand Scenario of Various Energy Items for Household Consumption in India under Various GDP Growth Rates

(Mtoe)

	Firewood and Chips		*Electricity*		*Dung Cake*		*Kerosene*		*LPG*	
Year	*8%*	*9%*	*8%*	*9%*	*8%*	*9%*	*8%*	*9%*	*8%*	*9%*
2000	79.62	79.62	8.43	8.43	29.61	29.61	10.07	10.07	6.42	6.42
2006	88.64	88.78	18.17	19.26	36.97	37.33	12.98	12.77	15.85	16.87
2011	94.11	94.05	27.17	29.68	40.42	40.48	14.01	14.02	23.94	26.07
2016	98.44	98.5	38.38	42.28	41.93	41.35	14.84	14.7	33.11	35.93
2021	102.06	102.46	50.39	54.78	41.79	40.87	15.16	14.93	41.63	44.16
2026	104.64	105.07	61.37	64.95	40.95	40.28	15.17	14.93	48.11	49.63
2031	106.39	106.59	69.72	71.8	40.47	40.21	15.12	14.96	52.27	52.89

Source: *TERI Energy Data Directory & Yearbook 2007.*

Table 3.8

The Impact of Electrification on the Demand Scenario of Various Energy Items for Household Consumption under Various GDP Growth Rates

(Mtoe)

	Firewood and Chips		*Electricity*		*Dung Cake*		*Kerosene*		*LPG*	
Year	*8%*	*9%*	*8%*	*9%*	*8%*	*9%*	*8%*	*9%*	*8%*	*9%*
2011	87.9	88	31.13	33.63	31.03	31.16	13.18	13.16	25.27	27.36
2016	92.59	93.02	42.58	46.51	32.21	31.53	13.82	13.64	34.3	36.95
2021	96.85	97.67	54.89	59.35	31.42	30.28	13.98	13.71	42.45	44.72
2026	1000.01	100.72	66.19	69.86	30	29.12	13.88	13.61	48.55	49.88
2031	102.08	102.41	74.82	76.95	29.14	28.78	13.76	13.59	52.49	53.05

Source: TERI Energy Data Directory & Yearbook 2007.

The above table shows that electrification will drastically curtail the use of the firewood and chips and to some extent the dung cakes and kerosene. LPG just has a positive association with the phenomenon.

Some Trends in the Electricity Sector

India now produces about 1,70,000 MW of electricity. However, not all the electricity reaches the consumer and hence, not all the demand for electricity is met by the national installed capacity.

Tables 3.9 and 3.10 show the power supply position, and the demand/supply gap in India's power sector since the beginning of Eleventh Plan until present.

Table 3.9

Power Supply Position since the Beginning of Ninth Plan until January 2010

Year	Requirement (MU)	Availability (MU)	Shortage (MU)	Shortage (%)
1997-98	424505	390330	34175	8.1
1998-99	446584	420235	26349	5.9
1999-2000	480430	450594	29836	6.2
2000-01	507216	467400	39816	7.8
2001-02	522537	483350	39187	7.5
2002-03	545983	497890	48093	8.8
2003-04	559264	519398	39866	7.1
2004-05	591373	548115	43258	7.3
2005-06	631554	578819	52735	8.4
2006-07	690587	624495	66092	9.6
2007-08	737052	664660	72392	9.8
2008-09	777039	691038	86001	11.1
2009-10*	688171	620003	68168	9.9

Note: * upto January, 2010 (data provision).

Source: Ministry of Power, Government of India, *Annual Report 2008-09*.

Table 3.10

Demand/Supply Gap in India's Power Sector

Year	Demand (MW)	Met (MW)	Shortage (MW)	Shortage (%)
1997-98	65435	58042	7393	11.3
1998-99	67905	58445	9460	13.9
1999-2000	72669	63691	8978	12.4
2000-01	78037	67880	10157	13
2001-02	78441	69189	9252	11.8
2002-03	81492	71547	9945	12.2
2003-04	84574	75066	9508	11.2
2004-05	87906	77652	10254	11.7
2005-06	93255	81792	11463	12.3
2006-07	100715	86818	13897	13.8
2007-08	108866	90793	18073	16.6
2008-09	109809	96785	13024	11.9
2009-10*	116281	101609	14672	12.6

Source: Same as Table 3.9.

Table 3.11

Region-wise Installed Capacity Statement

(All-India Installed Capacity (in MW) of Power Stations Located in the Regions of Mainland and Islands)

Region	Ownership Sector	Mode-wise Break-up							Grand Total
		Thermal			Total Thermal	Nuclear	Hydro (Renewable)	RES (MNRE)	
		Coal	Gas	Diesel					
Northern region	State	12394.50	1219.20	12.99	13626.69	0.00	7052.55	859.67	21538.91
	Private	135.00	0.00	0.00	135	0.00	786.00	1490.15	2411.15
	Central	8023.00	2344.06	0.00	10367.06	1180.00	5472.20	0.00	17019.26
	Subtotal	20552.50	3563.26	12.99	24128.75	1180.00	13310.75	2349.82	40969.32
Western region	State	15857.50	1804.72	17.28	17679.50	0.00	5480.50	338.88	23498.88
	Private	4180.00	2805.50	0.20	6985.70	0.00	447.00	4257.66	11690.36
	Central	6978.00	3533.59	0.00	10511.59	1840.00	1520.00	0.00	13871.59
	Subtotal	27015.00	8143.81	17.48	35176.79	1840.00	7447.50	4596.54	49060.83
Southern region	State	8822.50	555.70	362.52	9740.72	0.00	11107.03	939.89	21787.64
	Private	1110.00	3477.50	576.80	5164.30	0.00	0.00	6996.68	12160.98
	Central	7890.00	359.58	0.00	8249.58	1100.00	0.00	0.00	9349.58
	Subtotal	17822.50	4392.78	939.32	23154.60	1100.00	11107.03	7936.57	43298.20
Eastern region	State	6920.00	100.00	17.06	7037.06	0.00	3190.92	314.56	10542.54
	Private	1701.38	0.00	0.14	1701.52	0.00	0.00	20.20	1721.72
	Central	8024.00	90.00	0.00	8114.00	0.00	713.20	0.00	8827.20
	Subtotal	16645.38	190.00	17.20	16852.58	0.00	3904.12	334.76	21091.46

contd...

...contd...

Region	*Ownership Sector*	*Mode-wise Break-up*							*Grand Total*
		Thermal			*Total Thermal*	*Nuclear*	*Hydro (Renewable)*	*RES** (MNRE)*	
		Coal	*Gas*	*Diesel*					
Northeastern region	State	60.00	366.50	142.74	569.24	0.00	256.00	204.16	1029.40
	Private	0.00	24.50	0.00	24.50	0.00	0.00	0.00	24.50
	Central	0.00	375.00	0.00	375.00	0.00	860.00	0.00	1235.00
	Subtotal	60.00	766.00	142.74	968.74	0.00	1116.00	204.16	2288.90
Islands	State	0.00	0.00	50.02	50.02	0.00	0.00	5.25	55.27
	Private	0.00	0.00	20.00	20.00	0.00	0.00	0.00	20.00
	Central	0.00	0.00	0.00	0.00	0.00	0.00	0.00	0.00
	Subtotal	0.00	0.00	70.02	70.02	0.00	0.00	5.25	75.27
All-India	State	44054.50	4046.12	602.61	48703.23	0.00	27087.00	2662.41	78452.64
	Private	7126.38	6307.50	597.14	14031.02	0.00	1233.00	12764.69	28028.71
	Central	30915.00	6702.23	0.00	37617.23	4120.00	8565.40	0.00	50302.63
	Subtotal	82095.88	17055.85	1199.75	100351.48	4120.00	36885.40	15427.10	156783.98

Source: Same as Table 3.9.

Tables 3.12 and 3.13 show the extent of transmission and distribution losses, and the extent of demand/supply gap for power across various states in India.

Table 3.12

Transmission and Distribution Losses as a Percentage of Availability in States/Union Territories: 1997/98 to 2005/2006

State/Union Territory	*1997-98*	*1998-99*	*1999-2000*	*2000-01*	*2001-02*	*2002-03*	*2003-04*	*2004-05*	*2005-06*
State									
Andhra Pradesh	32.3	33.6	37.7	36.6	26.8	30.1	27.7	24	20.1
Arunachal Pradesh	34.4	30.6	47.1	34.4	53.6	39	47.5	43	49.7
Assam	27.3	38.7	39	40.7	42.8	38.3	39.3	51.8	40.3
Bihar	11	16.5	15.5	17.9	51.7	38	36.7	38.9	44
Chhattisgarh	—	—	—	—	33.8	37.9	42.6	28.1	31.1
Delhi	48.5	50.5	46.7	44.3	44	45.8	43.7	45.4	48.6
Goa	31	30.4	27.6	28.7	25.2	40.3	45.1	36	19.7
Gujarat	24.4	25.3	25.3	28.1	26.9	28.5	24.2	30.4	27.9
Haryana	34	35.3	38.3	39.8	39.2	37.7	32.1	32.1	30.5
Himachal Pradesh	20.2	26.3	27.8	23.4	25.6	21.25	22.8	28.9	23.6
Jammu and Kashmir	51.3	48	45	45.4	48.9	45.6	45.5	41.1	44.9
Jharkhand	—	—	—	—	26.4	21.2	25.4	19.6	26.8
Karnataka	19.1	30.6	37.3	34.9	33.8	24.6	23.3	26.1	29.8
Kerala	19.1	17.7	17.8	18.4	32.2	27.5	21.6	22.5	23.5
Madhya Pradesh	20.9	21.1	33.7	46.1	44.6	43.3	41.4	41.3	40.1
Maharashtra	18	17.8	29.2	33.8	37.3	34	34.1	32.4	31.6
Manipur	21	59.6	62.1	58.5	62.4	63.7	62.2	70.6	63.1
Meghalaya	12.1	19.9	27.4	21	22.7	21.9	16.7	28.4	40.2
Mizoram	46.8	44.8	47.6	45.4	49.8	46.9	55.5	66.1	39.2
Nagaland	29.8	26.5	32.3	24.6	52.3	26.7	55	48.3	59
Orissa	49.8	43.2	44.3	44.9	47.3	45.4	57.1	44	45.6
Punjab	18.8	18	18.3	56.6	27.7	24.4	26	25.4	27.6
Rajasthan	26.4	29.4	30.3	29.8	43.1	42.6	43.7	44.7	39.9
Sikkim	22.9	12.4	12.1	25	31.7	54.9	55	50.5	10.7
Tamil Nadu	17.1	16.8	17	15.7	16.1	17.3	17.2	19.3	18.7
Tripura	31.1	26.8	29.6	43.9	40.4	40.6	46.4	59.5	41.1
Uttar Pradesh	26.5	30.3	40.4	36.9	37.6	34.2	35.2	34.4	32.6
Uttarakhand	—	—	—	—	32.4	25.2	49.2	39.3	36
West Bengal	20.4	22.9	26.3	29.4	31.7	25.9	31	28.5	24.8

contd...

...contd...

State/Union Territory	*1997-98*	*1998-99*	*1999-2000*	*2000-01*	*2001-02*	*2002-03*	*2003-04*	*2004-05*	*2005-06*
Union territory									
Andaman and Nicobar Islands	20.6	20	16.5	17.5	29.2	19.8	26	12.6	24.3
Chandigarh	22.4	22.5	24.7	25.4	25	24.1	39.1	30.4	31.6
Dadra and Nagar Haveli	12.9	15.4	31.7	39.8	27.2	40.3	15.1	16	19.9
Daman and Diu	14.7	21.8	11.3	11.4	7.5	15	16.9	15.6	21.6
Lakshadweep	15.7	12.8	10.1	6.7	10.9	11.3	11.9	10.2	11.2
Puducherry	13.6	10.4	12.3	7.9	12	21.1	11.6	18.2	18.5
Damodar Valley Corporation	1.2	2.6	2.5	4.8	3.6	3.3	2.7	2.7	1.8
All-India	24.8	26.5	30.9	32.9	34	38.3	32.5	31.3	30.4

Source: *TERI Energy Data Directory & Yearbook 2007*.

The above table suggests that India loses over 30 per cent of electricity in the form of transmission and distribution losses. Only a few states *viz.*, Goa, Tamil Nadu, Sikkim and Andhra Pradesh have the transmission and distribution losses far below the national average in the year 2005-06.

Most of the states have a higher than national percentage. Incidentally, special regions like the Manipur, Assam, Jammu and Kashmir and even the national capital territory of Delhi have the losses far greater than the national average in the year 2005-06. Goa and Sikkim may be considered very small states and hence have lesser kilometres of power lines and hence are able to curtail the T&D losses. However, Delhi is also a small state. But, why should Delhi have a T&D losses of 48 per cent?

Table 3.13

Power Supply Position (2006/07): Surplus (+)/Deficit (-)

Region	Requirement (MU)	Availability (MU)	Surplus (+) Deficit (-) (MU)	(%)
Chandigarh	1,343	1,341	-2	-0.1
Delhi	22,397	22,012	-385	-1.7
Haryana	26,249	23,132	-3,117	-11.9
Himachal Pradesh	5,136	4,996	-140	-2.7
Jammu and Kashmir a	11,725	7,983	-3,742	-31.9
Punjab	38,641	34,839	-802	-9.8
Rajasthan	33,236	31,715	-1,521	-4.6
Uttar Pradesh	57,441	48,370	-9,071	-15.8
Uttarakhand	5,957	5,599	-358	-6
Northern region	202,125	179,987	-22,138	-11
Chhattisgarh	14,063	13,169	-894	-6.4
Gujarat	62,464	54,083	-8,381	-13.4
Madhya Pradesh	38,710	32,834	-5,876	-15.2
Maharashtra	110,005	89,138	-20,867	-19
Daman and Diu	1,602	1,408	-194	-12.1
Dadra and Nagar Haveli	2,923	2,879	-44	-1.5
Goa	2,624	2,606	-18	-0.7
Western region	232,391	196,117	-36,274	-15.6
Andhra Pradesh	60,964	58,280	-2,684	-4.4
Karnataka	40,797	39,948	-849	-2.1
Kerala	15,023	14,716	-307	-2
Tamil Nadu	61,499	60,445	-1,054	-1.7
Puducherry	1,808	1,808	0	0
Lakshadweep	25	25	0	0
Southern region	180,091	175,197	-4,894	-2.7
Bihar	8,425	7,741	-684	-8.1
Damodar Valley Corporation	11,542	11,308	-234	-2
Jharkhand	4,369	4,154	-215	-4.9
Orissa	17,101	16,796	-305	-1.8
West Bengal	26,538	25,966	-572	-2.2
Sikkim	223	218	-5	-2.2
Andaman and Nicobar Islands	240	190	-50	-20.8
Eastern region	68,198	66,183	-2,015	-3
Arunachal Pradesh	286	259	-27	-9.4
Assam	4,297	3,984	-313	-7.3
Manipur	451	429	-22	-4.9
Meghalaya	1,368	1,063	-305	-22.3
Mizoram	235	221	-14	-6
Nagaland	4,343	328	-15	-4.4
Tripura	802	728	-74	-9.2
Northeastern region	7,782	7,012	-770	-9.9
All-India	690,587	624,496	-66,091	-9.9

Source: *TERI Energy Data Directory & Yearbook 2007.*

The above table suggests that Chhattisgarh is the only state with a surplus peak time Power of over 6 per cent. While the national average peak time deficit is 9.9 per cent in 2006-07 which is about 15 per cent in 2010, it is a wonder as to why Jammu and Kashmir suffers from over 32 per cent peak time deficit in spite of abundant hydropower resources and lack of industrialisation.

Concluding Remarks

India has to go a long way to be fully recognised as the modern country. Still, 70 per cent of people cook using the firewood and chips and over 40 per cent of the rural household lack access to the electricity. While the modernisation efforts have to proceed unimpeded, the initiatives should be based on cleaner technologies like the renewable energy technologies so that the pollution can be controlled and sustainable development can be ensured.

India is a net importer of petroleum and natural gas. Heavy loss of foreign exchange and heavy subsidy burden are plaguing the Government.

Hence, at least the maintanence of the domestic production to import ratio is the paramount goal of the Government.

This reason has led to the introduction of NELP rounds leading to substantial increase in the domestic production of oil and natural gas and substantial increase in domestic reserves.

To earn valuable foreign exchange, India is actively pursuing the expansion of refineries and the export of refined products.

India has to ensure the penetration of electricity and electrical appliances while at the same time reducing deforestation and the use of kerosene. The transmission and distribution losses of the electricity should be controlled to a large extent and people and functionaries in the high loss states should be sensitised. Peak time management of electricity should be ensured using various physical and financial instruments of contract.

Appendix A-3.1

Model Production Sharing Contract (MPSC)

MPSC is the 'MoU' between the President of India, and the concerned company/companies involved in the exploration, development, discovery and production of oil and natural gas.

The MoU lists the functions, responsibilities and decision-making authority of the concerned stakeholders. MoU also discusses the operational aspects, general rights, financial burden to be shared, technical expertise to be used and to be procured through collaborations etc.

In so far as the petroleum is concerned, Articles 3, 4, 5 and 6 are crucial in the allotment of exploration blocks, rules for relinquishment of the exploration blocks, determination of the functions of the concerned company, and the formation and the constitution of the Management Committee. Article 10 relates to discovery, development and production. Article 21 pertains to natural gas.

Article-3: Licence and Exploration Period

3.1 The exploration period shall begin on the effective date and shall be for a period not exceeding seven (7) consecutive contract years in case of contract areas in onland and shallow water and not exceeding eight (8) consecutive contract years in case of deepwater area and frontier area and consisting of initial exploration period and subsequent exploration period. The initial exploration period shall consist of the first four (4) consecutive contract years in case of contract areas in onland and shallow water and first five (5) consecutive contract years in case of deepwater area and frontier area and with provision to proceed to the subsequent exploration period of maximum three consecutive years subject to provisions contained in Article 3.

3.2 Except as otherwise provided in this contract, the term of the initial exploration period shall not exceed the first four (4) consecutive contract years in case of contract areas in onland and shallow water and shall not exceed first five (5) consecutive contract years in case of deepwater areas and frontier areas in the exploration period.

3.3 Except as otherwise provided in this contract, the term of the subsequent exploration period shall not exceed three consecutive contract years from the end of the initial exploration period.

3.4 At the expiry of the initial exploration period, provided that the contractor has completed the minimum work programme, the contractor shall have the option, exercisable by giving a written notice to the Government at least thirty (30) days prior to the expiry of initial exploration period either:

a. to proceed to the subsequent exploration period by committing drilling of one exploratory well per year in contract area (in case of onland and shallow water blocks)/one exploratory well in three (3) years in contract area (in case of deepwater blocks) on presentation of the requisite guarantees as provided for in Article 29; in case of onland and shallow water blocks, the well(s) committed can be drilled in any of the years, as the case may be, of the subsequent exploration period. In case of deepwater blocks the well(s) committed can be drilled in any of the three years of the subsequent exploration period. or

b. to relinquish the entire contract area except for any discovery area and any development area and to conduct development operations and production operations in relation to any commercial discovery in accordance with the terms of this contract, and the contractor shall have no further obligation in respect of the minimum work programme under Article 5.

If neither of the options provided for in paragraphs (a) and (b) hereof is exercised by the contractor, this contract shall terminate at the end of initial exploration period and the licence shall be automatically cancelled.

3.5 If at the end of the initial exploration period the minimum work programme for that period is not completed, the time for completion of the said minimum work programme shall be extended for a period necessary to enable completion thereof but not exceeding six (6) months, provided that the contractor submits his request by giving a written notice to the Government at least thirty (30) days prior to the expiry of the initial exploration period and can show technical or other good reasons for non-completion of the minimum work programme and the management committee gives its prior written consent to the said extension and provided further that the period of such extension shall be subtracted from the succeeding subsequent exploration period. In case the minimum work programme of the initial exploration period is completed before stipulated time as provided in the Article 3.2, the time so saved will be added to the succeeding subsequent exploration period, if so requested by the contractor giving a notice in writing to the Government thirty (30) days prior to such early completion of the initial exploration period and in that event the provision of the Article 3.4 (a) shall apply immediately after such early completion of initial exploration period.

3.6 If, at the end of the initial exploration period, execution of any work programme is in progress and which is in addition to the minimum work programme, such initial exploration period shall be extended for a period not exceeding six (6) months to enable completion thereof provided that the minimum work programme for such period has been completed and the management committee gives its prior written consent to the said extension as provided in the Article 3.5. In the event of an extension as provided for herein, the notice referred to in Article 3.4 shall be given at least thirty (30) days prior to the expiry of the relevant extension.

3.7 Where sufficient time is not available prior to the expiry of the exploration period to complete an appraisal programme, at the request of the contractor, the Government shall extend the exploration period for such period, not exceeding eighteen (18) months for onland and shallow water blocks and thirty (30) months for deepwater blocks, as may be mutually agreed between the parties for the appraisal programme to be carried out and for the contractor and the management committee, to comply with the provisions of Article 10 and Article 21.

3.8 If no commercial discovery has been made in the contract area by the end of the exploration period, the contract shall terminate.

3.9 If this contract is terminated in accordance with its terms, the licence shall be automatically cancelled.

3.10 If at the expiry of the exploration period a development plan for development of a commercial discovery and an application for lease is under consideration by the management committee or Government, as the case may be, pursuant to Articles 10, 11 and 21 respectively, the licence shall continue in force with respect to that part of the contract area to which the application for the lease relates, pending a decision on the proposed development plan and the application for the lease, but shall cease to be in force and effect with respect to the remainder of the contract area.

Article-4: Relinquishment

4.1 At the end of the initial exploration period, the contractor shall have option to relinquish contract area after completion of minimum work programme or proceed to the subsequent exploration period and retain the contract area by committing to carry out drilling of one well per year in contract area (in case of onland and shallow water blocks)/one well in three years in contract area (in case of deepwater blocks).

The contract area (excluding discovery and development area) shall be relinquished at the end of 7 or 8 consecutive years, as the case may be, of exploration period.

4.2 At the end of the exploration period, the contractor shall retain only development areas and discovery areas.

4.3 If the contractor exercises the option provided for in paragraph (b) of Article 3.4, the contractor shall, after any discovery areas or development areas have been designated, relinquish all of the contract area not included within the said discovery areas or development areas.

4.4 As and when the contract is terminated under the provisions of Article 3 or in accordance with any other provisions of this contract, the entire contract area remaining with the contractor shall be deemed to have been relinquished by the contractor as on the date on which the contract is terminated.

4.5 Relinquishment of all or part of the contract area or termination of the contract shall not be construed as absolving the contractor of any liability undertaken or incurred by the contractor in respect of the contract area during the period between the effective date and the date of such relinquishment or termination.

4.6 Subject to Article 14.9, the liability of the contractor shall be limited to any liability undertaken or incurred in respect of, relating to or connected with the contract, and/or any claim arising out of or in relation to the act of negligence, misconduct, commission or omission in carrying out petroleum operations during the period between the effective date and the date of relinquishment of the contract area or termination or expiry of the contract, as the case may be.

Article-5: Work Programme

5.1 The contractor shall commence petroleum operations not later than six (6) months from the effective date.

5.2 In addition to the work programme commitment in initial exploration period specified below in Article 5.2.1 and 5.3, the contractor shall be required to undertake and complete the 2D seismic API, in grid size of ___ LKM x ___ LKM covering the entire contract area (herein after referred to as 'mandatory work programme') during initial exploration period. In case due to any reason intrinsic to the contract area, the contractor is not able to cover any part of the contract area by 2D seismic survey of grid size specified in this Article, the contractor shall submit a proposal for substitution of the shortfall in the mandatory work programme to the management committee for approval. The management committee shall consider and take a reasoned unanimous decision on the proposal of the contractor in a timely manner.

The management committee may ask the contractor by giving a reasonable time, any relevant information/details/data to enable it to take a decision on the proposal. The management committee shall ensure that the substituted work programme shall be at least equal to the shortfall in the mandatory work programme when evaluated in terms of efforts and expenditure. In case management committee is not able to decide unanimously, the matter may be referred for approval of the Government. In case no proposal is received from the contractor for the substitution and fails to complete mandatory work programme the provision of Article 5.6 shall apply. However, if the work programme of 3D seismic API in Article 5.2.1 (a) is equal in size to the contract area then the contractor shall be exempted from carrying out the 2D seismic mandatory work programme.

5.2.1 During the currency of the initial exploration period, as per Article 3.2, the contractor shall complete the following work programme:

(a) a seismic programme consisting of the acquisition, processing and interpretation of line kilometres of 2D and sq kms of 3D seismic data in relation to the exploration objectives; and

(b) ____ exploration well(s) shall be drilled to at least one of the following depths:

(i) ____ metres (from sea bed);

(ii) to basement and

(iii) that point below which further drilling becomes impracticable due to geological conditions encountered and drilling would be abandoned by a reasonable prudent operator in the same or similar circumstances. Abandonment of drilling under this provision by the contractor would require unanimous approval of the management committee.

Note: The details of the work programme to be filled in the Article 5.2 & 5.2.1 shall be as per the NIO and the bid of company(ies) and accepted by the Government.

5.3 During the currency of subsequent exploration period, if opted in accordance with Article 3.4 (a), the contractor shall complete the following work programme:

For onland blocks and shallow water blocks: One (1) exploration well in each year

or

For deepwater blocks: one (1) exploration well

to at least one of the following depths:

(i) ____metres from sea bed; (to be specified by the contractor at the time of exercising the option)

(ii) to Basement; and

(iii) that point below which further drilling becomes impracticable due to geological conditions encountered and drilling would be abandoned by a reasonable prudent operator in the same or similar circumstances. Abandonment of drilling under this provision by the contractor would require unanimous approval of the management committee.

5.4 The contractor shall ensure that all relevant subsurface, geological, geochemical and geophysical information necessary for the attainment of the exploration objectives in accordance with modern oilfield and petroleum industry practices is obtained during exploratory drilling.

5.5 If the depth of the well is not achieved for any reason, a substitute Well shall be drilled of the same specifications as stipulated in and subject to Articles 5.2 & 5.3, as the case may be.

5.6 Subject to Article 31, the contractor undertakes to complete the mandatory work programme and minimum work programme in accordance with Articles 5.2, 5.2.1, 5.3 and 5.5, as the case may be. In the event that the contractor fails to fulfil the said mandatory work programme or minimum work programme or work programme committed during the initial exploration period or subsequent exploration period or early termination of the contract by the Government for any reason whatsoever, each company constituting the contractor shall pay to the Government, within sixty (60) days following the end of the initial exploration period or subsequent exploration period or early termination of the contract, as may be the case, its participating interest share for an amount which shall be equivalent to liquidated damages as specified below:

(In US$)

	Onland	*Shallow Water*	*Deepwater*
Per well	1,000,000	3,000,000	6,000,000
Per sq km of 3D	5,000	1,500	1,500
Per line km of 2D	2,500	1,000	1,000

In case of 2D seismic and 3D seismic, the contractor shall pay LD for the quantum of work programme that falls short of the minimum work programme. However, in case of well(s) which have not been drilled to the depth specified in the minimum work programme, the contractor shall pay LD for the entire well, irrespective of the metreage left to be drilled.

5.7 If the work programme for the subsequent exploration period has been completed earlier than specified period, the contractor shall meet with the Government to discuss the possibility of early relinquishment, unless the contractor undertakes further work with the approval of the management committee.

5.8 As soon as possible after the effective date and thereafter within ninety (90) days before commencement of each following year, the contractor shall submit to the management committee the work programmes and the budgets relating to petroleum operations to be carried out during the relevant year. Work programme and budgets for the exploration period shall include work sufficient to meet the relevant mandatory work programme and minimum work programme with respect to initial exploration period specified in this Article 5.

5.9 The contractor may propose modifications or revisions to the details of a reviewed or an approved work programme and budget, as the case may be, in the light of the then existing circumstances and shall submit to the management committee modifications or revisions to the work programme and budget referred to in Article 5.8

5.10 Work programmes and budgets and any modifications or revisions thereto relating to exploration operations shall be submitted to the management committee for review and advice as provided in Article

Article-6: Management Committee

6.1 There shall be constituted a committee to be called the management committee with functions as stated herein below.

6.2 Government shall nominate two (2) members representing Government in the management committee, whereas each company constituting the contractor shall nominate one (1) member each to represent company in the management committee provided that in case the contractor constitutes only one company, that company shall have two (2) members. The parties shall nominate the members to the management committee within thirty (30) days of the effective date.

6.3 Each party may nominate alternate members with full authority to act in the absence and on behalf of the members nominated under Article 6.2 and may, at any time, nominate another member or alternate member to replace any member nominated earlier by notice to other members of the management committee.

6.4 One representative of the Government shall be designated as the chairman of the management committee and the second representative of the Government shall be designated as the deputy chairman. The member of the operator, or the member designated by the operator where operator has two (2) members in the management committee shall be designated as the secretary of the committee.

6.5 Operator on behalf of the contractor with the approval of operating committee, if constituted under the Article 7.4, or in case of a single party constituting the contractor, then that party shall submit following matters to the management committee for review and it shall have advisory functions:

(a) prior to the date of declaration of first commercial discovery, the annual work programmes and budgets in respect of exploration operations and any revisions or modifications thereto:

(aa) after the date of declaration of first commercial discovery, the annual work programme and budgets in respect of exploration operations relating to minimum work programme during initial exploration period and additional committed work programme during subsequent exploration period, if opted under Article 3.4, and any revisions and modifications thereto;

(b) annual work progress and costs incurred thereon;

(c) proposals for surrender or relinquishment of any part of the contract area by the contractor;

(d) proposals for an appraisal programme or revisions or additions thereto and the declaration of a discovery as a commercial discovery;

(e) any other matter required by the terms of this contract to be submitted to it for review or advice; and

(f) any other matter which the contractor decides to submit for review or advice including matters concerning inter-party relationships.

6.6 The following matters shall be submitted by operator on behalf of the contractor with the approval of operating committee, if constituted under the Article 7.4, or in case of single party constituting the contractor, then by that party to the management committee for approval:

(a) except for items specified in 6.5 (a) and 6.5 (aa), annual work programme and budgets in respect of exploration operations and any modifications and revisions thereto;

(aa) annual work programme and budgets in respect of development operations and production operations and any modifications or revisions thereto;

(b) proposals for the approval of development plans as may be required under this contract, or modifications or revisions to a development plan;

(c) determination of a development area;

(d) appointment of auditors along with scope of audit, approval and adoption of audited report submitted under Article 25.4.3;

(e) collaboration with licensees or contractors of other areas;

(f) claims or settlement of claims for or on behalf of or against the contractor in excess of limits fixed by the management committee from time to time;

(g) proposal about abandonment plan/site restoration as required to be submitted under Article 14.10;

(h) any other matter required by the terms of this contract to be submitted for the approval of the management committee;

(i) any other matter which the contractor decides to submit to it; and

(j) any matter, which Government refers to the management committee for its consideration and reasoned opinion.

6.7 Unless agreed otherwise by all the members of the management committee, the management committee shall meet at least once every six (6) months during the exploration period and thereafter at least once every three (3) months or more frequently at the request of any member. The secretary, with the approval of the chairman, shall convene each meeting by notifying the members twenty-eight (28) days prior to such a meeting (or a shorter period of notice if the members unanimously so agree) of the time and place of such meeting and the purpose thereof and shall include in such notice a provisional agenda for such meeting.

The chairman shall be responsible for processing the final agenda for such meeting and the agenda shall include all items of business requested by the members to be included, provided such requests are received by the secretary at least ten (10) days prior to the date fixed for the meeting. The secretary shall forward the agenda to the members at least seven (7) business days prior to the date fixed for the meeting. Matters not included in the agenda may be taken up at the meeting by any member with the unanimous prior written consent of all the members whether present or not present at the meeting.

6.8 The chairman or the deputy chairman, as may be the case, shall preside over the meetings of the management committee and, in their absence, any other member representing Government and present shall preside over the meetings.

6.9 Secretary to the management committee shall be responsible, *inter alia*, for preparation of the minutes of every meeting in the English language and provision to every member of the management committee with two (2) copies of the minutes approved by the chairman within three (3) business days of the meeting. Unless agreed otherwise by all the members of the management committee, the minutes of a meeting shall be finalised by the management committee within three (3) business days thereafter.

Members shall notify the chairman and the other members of their approval of the minutes by putting their signatures on one copy of the minutes and returning the same to the chairman. Members may suggest any modification to the minutes while returning the signed copy. Members may also communicate with the chairman through telex, cable, or facsimile or any other effective mode of communication agreed by all the members of the management committee. if the chairman or any other member does not agree with the modification to the minutes suggested by any member, the matter shall be brought to the attention of the other members and resubmitted to the management committee at the next meeting and the minutes shall stand approved as to all other matters. If a member fails to respond within the aforesaid three (3) business day period, unless agreed otherwise by the management committee as herein provided, the minutes shall be deemed to be approved by such member.

6.10 Any member shall be entitled, if either he/she or his/her alternate is unable to attend a meeting, to cast his vote by telex, cable, facsimile transmission or any other effective mode of communication agreed by all the members of the management committee and received by the chairman prior to the date on which the vote is taken in the course of the meeting or by giving a prior written notice to all other members, appoint a member, with his/her prior written consent, representing another party in the management committee as its proxy to attend a meeting and to exercise the appointing member's right to vote at the meeting whether as directed by the appointing member or otherwise. A member appointed as a proxy and attending a meeting shall be present in two separate capacities and vote accordingly. All such votes shall have the same effect as if that member had been present and so voted at the meeting.

6.11 In case of urgency, where operating committee has made a recommendation together with reasons to the chairman requiring consideration of a matter by the management committee without delay, chairman, after being satisfied may waive the requirements of notice period for the meeting and circulation of agenda to such extent as would be consistent with the urgency and consideration of the matter by the management committee. Alternatively, chairman may approve submission of notice and agenda to members by telex or facsimile transmission or any other effective mode of communication agreed by all the members of the management committee, receipt of which shall be confirmed by telephone by the chairman requiring the members to confirm their decision by these modes of communication not later than three (3) business days from confirmation of receipt of notice and agenda by the member. Any member failing to convey the decision within the time limits of three (3) business days shall be deemed to have voted in favour of the proposal. The result of any such vote shall be notified by the chairman to all the members.

6.12 The meetings of the management committee shall be held in India, unless otherwise mutually agreed by the members of the management committee. All expenses of the members of the management committee attending meetings shall be borne by the respective party and shall in no event be cost recoverable.

6.13 All matters requiring the approval of the management committee shall be generally approved by a unanimous vote of the members of the management committee present as well as the views of the members received by some other mode of communication. In case, unanimity is not achieved in decision-making process within a reasonable period as may be required under the circumstances, the decision of the management committee shall be approved by the majority participating interest of seventy per cent (70%) or more with Government representative having a positive vote in favour of the decision.

6.14 There shall be a quorum of the management committee for holding a meeting and making decisions with each party to the contract represented by at least one of its nominated members in the management committee either present in person or represented as per Article 6.10. If there is no quorum in a meeting, the meeting shall stand postponed to the same day and time in the next week and if quorum is not present or represented even in the next meeting and subject to a Government member being present, the members present and represented will constitute the quorum and take decisions and decisions taken by such quorum shall be final and binding to all the absenting parties or parties not represented, notwithstanding the provisions of Article 6.13.

6.15 The management committee, if it considers necessary, may appoint legal, financial or technical subcommittees comprising such representatives as may be agreed by the management committee to consider any matter requiring approval or decision of the management committee. Such subcommittee expenses shall form part of contract cost with relevant cost classification as decided by the management committee pursuant to the Section 2 of the Accounting Procedure and will be cost recoverable.

6.16 In the event a party to the contract is not entitled to vote in the operating committee meetings being in default under the operating agreement, and operator notifies chairman of the default by the party, then the issue of exercising voting right by such defaulting party in the management committee meetings shall be discussed by the management committee. The management committee excluding the defaulting party, after duly hearing the views of the defaulting party on the matter of their default under operating agreement, shall take unanimous decision on exclusion or otherwise of the defaulting party from voting in the management committee meetings.

For avoidance of any doubt, it is clearly understood that unanimous decision by the management committee referred to in this Article 6.16 excludes defaulting party from such decision. Accordingly, if the management committee decides to exclude the defaulting party from voting in the management committee, then the said party shall not be entitled to vote in the meetings of the management committee under contract. In that event, for the purpose of the provisions of Article 6.13, decisions of the management committee shall be made by vote of the members of the management committee excluding the member appointed by the said party in default and any vote or purported vote by such member in the management committee shall be ignored. The said party in default shall be bound by all decisions of the management committee. The non-defaulting parties under the operating agreement shall indemnify Government against any claims of whatsoever nature which may arise due to exclusion of defaulting party from voting in the management committee.

Article-10: Discovery, Development and Production

10.1 If and when a discovery is made within the contract area, the contractor shall:

a. forthwith inform the management committee and Government of the discovery;

b. promptly thereafter, but in no event later than a period of thirty (30) days from the date of the discovery, furnish to the management committee and Government particulars, in writing, of the discovery; and

c. run tests promptly and in any case within 90 days from the date under Article 10.1 (a) to determine whether the discovery is of potential commercial interest and, within a period of sixty (60) days after completion of such tests, submit a report to the management committee containing data obtained from such tests and its analysis and interpretation thereof, together with a written notification of whether, in the contractor's opinion, such discovery is of potential commercial interest and merits appraisal.

10.2 If the contractor determines to conduct a drill stem or production test, in open hole or through perforated casing, with regard to any exploration well, it shall notify the Government of the time of such test at least forty-eight (48) hours prior to the proposed test, and the Government shall have the right to have a representative present during such test.

10.3 If, pursuant to Article 10.1 (c), the contractor notifies the management committee that the discovery is of potential commercial interest, the contractor shall prepare and submit to the management committee within one hundred and twenty (120) days of such notification, a proposed appraisal programme with a work programme and budget to carry out an adequate and effective appraisal of such discovery designed to achieve both the following objectives: (i) determine without delay, and, in any event, within the period specified in Article 10.5, whether such discovery is a commercial discovery and (ii) determine, with reasonable precision, the boundaries of the area to be delineated as the development area.

10.4 The proposed appraisal programme shall be reviewed by the management committee within thirty (30) days after submission thereof pursuant to Article 10.3. The said appraisal programme, together with the work programme and budget submitted by the contractor, which may be revised or modified or amended by the contractor in light of the management committee review, shall be adopted as the appraisal programme and the contractor shall promptly commence implementation thereof; and the annual budget for the exploration period, adopted pursuant to Article 5, shall be revised accordingly.

10.5 The contractor shall in respect of a discovery of crude oil, advise the management committee by notice in writing within a period of eighteen (18) months for onland and shallow water blocks and thirty (30) months for deepwater blocks from the date on which the notice provided for in Article 10.1 (c) was delivered, whether such discovery should be declared a commercial discovery or not. Such notice shall be accompanied by a report on the discovery setting forth all relevant technical and economic data including estimated recoverable reserves, sustainable production levels, estimated development and production expenditures, prevailing and forecasted prices, and other pertinent technical and economic factors according to modern oilfield and petroleum industry practices as well as all evaluations, interpretations and analyses of such data and feasibility studies relating to the discovery prepared by or for the contractor, with respect to the discovery and any other relevant information. If the contractor is of the opinion that crude oil has been discovered in commercial quantities, it shall submit the proposal to the management committee for review that the discovery be declared a commercial discovery. In the case of a discovery of gas, the provisions of Article 21 shall apply.

10.6 The management committee shall, within forty (40) days of the date of the notice referred to in Article 10.5, review the proposal of the contractor and request any other additional information it may reasonably require so as to complete the review of the proposal made by the contractor. The contractor shall furnish the additional information within thirty (30) days from the date of the request. The review by the management committee shall be made and conveyed to the contractor within the later of: (a) ninety (90) days from the date of notice referred to in Article 10.5 or (b) forty (40) days of receipt of such other information as may be required under this Article.

10.7 If the contractor declares the discovery a commercial discovery after taking into account the advice of the management committee as referred in the Article 10.6, within two hundred (200) days of the declaration of the discovery as a commercial discovery, the contractor shall submit to the management committee a comprehensive development plan of the commercial discovery which shall:

a. relate to the discovery area and contain a reservoir or part thereof and the boundaries of the proposed development area;
b. be designed to ensure the most efficient, beneficial and timely use of the petroleum resources discovered; and
c. be prepared in accordance with sound engineering, economic, safety and environmental principles recognised in the generally accepted modern oilfield and petroleum industry practices.

Such plan shall contain detailed proposals by the contractor for the construction, establishment and operation of all facilities and services for and incidental to the recovery, storage and transportation of the petroleum from the proposed development area to the delivery point together with all data and supporting information including but not limited to:

- description of the nature and characteristic of the reservoir, data, statistics, interpretations and conclusions on all aspects of the geology, reservoir evaluation, petroleum engineering factors, reservoir models, estimates of reserve in place, possible production magnitude, nature and ratio of petroleum fluids and analysis of producible petroleum;
- outlines of the development project and/or alternative development projects, if any, describing the production facilities to be installed and the number of wells to be drilled under such development project and/or alternative development projects, if any;

- estimate of the rate of production to be established and projection of the possible sustained rate of production in accordance with modern oilfield and petroleum industry practices under such development project and/or alternative development projects, if any, which will ensure that the area does not suffer an excessive rate of decline of production or an excessive loss of reservoir pressure;
- estimates of development costs and production costs under such development project and/or alternative development projects, if any;
- Contractor's recommendations as to the particular project that it would prefer;
- Work programme and budget for development proposals relating to the proposed development area;
- anticipated adverse impact on the environment and measures to be taken for prevention or minimisation thereof and for general protection of the environment in conduct of operations;
- measures to be taken for the health and safety of persons employed in petroleum operations;
- the information required in Article 21.
- schedule of implementation and targets

10.8 A proposed development plan submitted by the contractor pursuant to Article 10.7 may be approved by the management committee within one hundred and ten (110) days of submission thereof or eighty (80) days of receipt of any additional information requested by the management committee. In case the management committee requires any reasonable additional information, the same shall be requested by it within eighty (80) days from the submission of the development plan. The contractor shall provide such additional information within thirty (30) days from the request by the management committee. If, within a period of one hundred and ten (110) days after submission of a proposed development plan or eighty (80) days from the receipt of any additional information, where asked by the management committee, the management committee fails to convey a decision to the contractor, the contractor shall have option to submit the proposal to the government. Also, where, the management committee rejects the development plan of the contractor, the contractor can submit the development plan for the approval of the government.

The Government shall respond on the proposed development plan submitted by the contractor within one hundred and ten (110) days. In case Government refuses to approve the proposed development plan, it shall convey the reasons for such refusal and the contractor shall be given opportunity to make appropriate modifications to meet concerns of Government and the provisions of the foregoing Article and re-submit the plan within ninety (90) days from the date of receipt of refusal from the Government.

10.9 A development plan approved by the management committee or Government, as may be the case, from time to time shall commit the contractor to the obligations stipulated in Articles 10.10 to 10.12.

10.10 Work programmes and budgets for development and production operations shall be submitted to the management committee as soon as possible after the approval of a development plan under Article 10.8 29 and thereafter not later than 31st December each year in respect of the year immediately following.

10.11 The management committee, when considering any work programme and budget, may require the contractor to prepare an estimate of potential production to be achieved through the implementation of the said work programme and budget for each of the three (3) years following the year to which the work programme and budget relate. If major changes in yearly estimates of potential production are required, these shall be based on evidence necessitating such changes.

10.12 Not later than the fifteenth (15th) of January each year, in respect of the year immediately following commencement of commercial production, the contractor shall determine the 'programme quantity' with the approval of the management committee. The programme quantity for any year shall be the maximum quantity of petroleum based on contractor's estimates, as approved by the management committee, which can be produced from a development area consistent with modern oilfield and petroleum industry practices and minimising unit production cost, taking into account the capacity of the producing wells, gathering lines, separators, storage capacity and other production facilities available for use during the relevant year, as well as the transportation facilities up to the delivery point.

10.13 Proposed revisions to the details of a development plan or an annual work programme or budget in respect of development and production operations shall, for good cause and if the circumstances so justify, be submitted for approval to the management committee.

10.14 In the event the area encompassing the commercial discovery extends beyond the development area designated in the development plan, either within the original contract area but subsequently relinquished or, outside the original contract area, the management committee may make recommendations to the Government concerning enlargement of the development area, provided the same was not awarded to any other company by the Government or is not held by any other party or not on offer by the Government and no application for a licence or lease is pending with the Government. However, in case the area is held by any other party or on offer by the Government or application for licence or lease is pending with the Government, the management committee shall notify the same to the Government for further action on the matter. Government may consider such request for extension at its sole discretion and on terms and conditions, which it may consider fit.

Article-21: Natural Gas

21.1 Until such time as the total availability of natural gas from all petroleum production activities in India meets the total national demand as determined by the Government, each company comprising the contractor, shall sell in the domestic market in India all of the company's entitlement to natural gas from the contract area.

21.1.1 If India attains self-sufficiency in natural gas, during any year, the Government shall advise the company (ies) accordingly by a written notice. In such an event, domestic sale obligation shall be suspended for such period as may be specified by the Government, and the company shall have the right to lift and export its participating interest share of natural gas during the said period, subject to any other extant policy guidelines of the Government applicable from time to time.

21.1.2 If self-sufficiency ceases to exist, the position shall revert to domestic sale obligation as outlined in Article 21.1.

21.2 The contractor shall have the right to use natural gas produced from the contract area for the purpose of petroleum operations including reinjection for pressure maintenance in oilfields, gas lifting and captive power generation required for petroleum operations.

21.3 For the purpose of domestic sale obligation, the contractor shall have freedom to market the natural gas and sell its entitlement as per Government policy for utilisation of gas among different sectors.

21.3.1 For purposes of Article 21, the Government may from time to time frame policy for utilisation of gas among different sectors, both for associated natural gas (ANG) as well as non-associated natural gas (NANG) which would cover issues relating to gas supplies to different consumer sectors.

21.4 Associated natural gas (ANG):

21.4.1 In the event that a discovery of crude oil contains ANG, the contractor shall declare in the proposal for the declaration of the said discovery as a commercial discovery as specified in Article 10, whether (and by what amount) the estimated production of ANG is anticipated to exceed the quantities of ANG which will be used in accordance with Article 21.2 (such excess being hereinafter referred to as 'the excess ANG'). In such an event the contractor shall indicate whether, on the basis of the available data and information, it has reasonable grounds for believing that the excess ANG could be commercially exploited in accordance with the terms of this contract along with the commercial production of the crude oil from the contract area, and whether the contractor intends to so exploit the excess ANG.

21.4.2 Based on the principle of full utilisation and minimum flaring of ANG, a proposed development plan for an oil discovery shall, to the extent practicable, include a plan for utilisation of the ANG including estimated quantities to be flared, reinjected, and to be used for petroleum operations; and, if the contractor proposes to commercially exploit the excess ANG for sale in the domestic market in accordance with Government's policy, or elsewhere, the proposed plans for such exploitation.

21.4.3 If the contractor wishes to exploit the excess ANG, subject to Article 21.1, the contractor shall be free to explore markets for the commercial exploitation of the said excess ANG and submit its proposals for such exploitation to the Government in accordance with Article 21.4.2.

21.4.4 Where the contractor is of the view that the excess ANG cannot be commercially exploited and chooses not to exploit the said excess ANG, or is unable to find a market for the Excess ANG pursuant to Article 21.4.3, the Government shall be entitled to take and utilise such excess ANG free of any cost/ charge.

21.4.5 If the Government elects to take the excess ANG as provided in Article 21.4.4:

a. the contractor shall deliver such excess ANG to the Government (or its nominee) free of any cost/charge, at the downstream flange of the gas/oil separation facilities;
b. the contractor shall, based on sound petroleum engineering practices, install such facilities as would facilitate, insofar as practicable, uninterrupted delivery of such excess ANG to the Government or its nominee;

c. the cost of all facilities installed pursuant to paragraph (b) above shall be borne by the Government (or its nominee);

d. the Government or its nominee shall bear all costs including gathering, treating, processing and transporting costs beyond the downstream flange of the gas/oil separation facilities; and

e. the delivery of such excess ANG shall be subject to procedures to be agreed between the Government or its nominee and the contractor prior to such delivery, such procedures to include matters relating to timing of off-take of such excess ANG. Parties shall endeavour that such procedures do not restrict oil production.

21.4.6 The excess ANG which is not commercially exploited by the contractor, or taken by the Government or its nominee pursuant to this Article 21, shall be returned to the subsurface structure or flared or otherwise disposed off as approved by the Government in the context of the development plan, provided that flaring will be resorted to only for small quantities and as a last resort.

21.4.7 As soon as practicable after the submission of the proposed development plan, the contractor and the Government or its nominee shall meet to discuss the sale and/or disposal of any ANG discovered with a view to giving effect to the provisions of this Article 21 in a timely manner.

21.5 *Non-associated natural gas (NANG)*:

21.5.1 In the event of a discovery of NANG in the contract area, the contractor shall promptly report such discovery to the management committee and the Government and the provisions of Articles 10.1 and 10.2 shall apply. The remaining provisions of Article 10 would apply to the discovery and development of NANG only insofar as they are not inconsistent with the provisions of this Article. Notwithstanding the provisions of Article 3, the contractor shall be entitled to retain the discovery area subject to the provisions of this Article 21.

21.5.2 If, pursuant to Article 10.1, the contractor gives notification that the discovery is of potential commercial interest, the contractor shall submit to the management committee, within one (1) year from the date of notification of the above said discovery, the proposed appraisal programme, including a work programme and budget to carry out an adequate and effective appraisal of such discovery, to determine: (i) without delay, whether such discovery is a commercial discovery and (ii) with reasonable precision, the boundaries of the area to be delineated as the development area. Such proposed appraisal programme shall be supported by all relevant data such as well data, contractor's best estimate of reserve range and production potential, and shall indicate the date of commencement of the proposed appraisal programme.

21.5.3 The proposed appraisal programme together with the work programme and budget referred to in Article 21.5.2 shall be reviewed by the management committee within sixty (60) days of its submission by the contractor. The management committee shall offer its comments within the said period. The said appraisal programme together with the Work programme and budget submitted by the contractor as revised or modified or amended in light of the management committee review and advice, shall be adopted as the appraisal programme and the contractor shall promptly proceed with implementation of the said programme.

21.5.4 If on the basis of the results of the appraisal programme, the contractor is of the opinion that NANG has been discovered in commercial quantities, it shall submit to the management committee as soon as practicable but not later than three (3) years from the date of notification of the aforementioned discovery, a proposal for the declaration of the discovery as a commercial discovery. Such proposal shall take into account the Government's policies on gas utilisation and propose alternative options, if any for use or consumption of the NANG and be accompanied by a report on the discovery supported by, *inter alia*, technical and economic data, evaluations, interpretations and analyses of such data and feasibility studies relating to the discovery prepared by or on behalf of the contractor, and other relevant information. If no proposal is submitted to the management committee by the contractor within three (3) years from the said discovery, the contractor shall relinquish its rights to develop such discovery and the area relating to such discovery shall be excluded from the contract area.

21.5.5 Where the contractor has submitted a proposal for the declaration of a discovery as a commercial discovery, the management committee shall consider the proposal of the contractor with reference to commercial utilisation or commercial development of the NANG in the domestic market or elsewhere and in the context of Government's policy on gas utilisation and the chain of activities required to bring the NANG from the delivery point to potential consumers in the domestic market or elsewhere. The management committee may, within eighty-five (85) days of the submission of the said proposal, request the contractor to submit any additional information on the discovery, the anticipated markets or any other related matter that may reasonably be required to facilitate a review. The contractor shall submit the required information within thirty (30) days of the request by the management committee. The management committee will advise the contractor of its review within one hundred and thirty-five (135) days from the submission of proposal or within fifty-five (55) days from the receipt of additional information, as the case may be, on the proposal made by the contractor to declare the discovery as a commercial discovery.

21.5.6 If the contractor declares the discovery a commercial discovery after taking into account the advice of the management committee as referred to in the Article 21.5.5, the contractor shall, within one (1) year of the declaration of the discovery as a commercial discovery, submit a development plan for the development of the discovery to the management committee for approval. Such plan shall be supported by all relevant information including, *inter alia*, the information required in Article 10.7.

21.5.7 Unless otherwise agreed by the management committee, it shall consider the proposed development plan and give their approval within one hundred and sixty-five (165) days of submission thereof or eighty-five (85) days from the receipt of the clarifications/additional information from the contractor. Any clarification/additional information required by the management committee shall be asked for within eighty-five (85) days of receipt of the proposal from the contractor. The contractor shall provide such additional information within thirty (30) days from the receipt of request by the management committee. If the management committee fails to convey its decision within one hundred and sixty-five (165) days from the submission of the development plan or eighty-five (85) days from the receipt of the clarifications/additional information, whichever is later, the contractor may submit the development plan for the approval of the Government. Also, where, the management committee rejects the development plan of the contractor, the contractor can submit the development plan for the approval of the Government.

21.5.8 Where the development plan is submitted to the Government for approval pursuant to Article 21.5.7, the Government shall convey its decision within one hundred and fifteen (115) days from the date of receipt of the proposal from the contractor. Government, where it considers necessary, may ask clarifications/additional information from the contractor within eighty-five (85) days and shall convey its decision within fifty-five (55) days from the date of receipt of such clarifications/additional information.

21.5.9 If the Government has failed to approve or disapproves the contractor's proposed development plan, within one hundred and fifteen (115) days from receipt or within fifty-five (55) days from the receipt of clarifications/ information from the contractor as mentioned in the Article 21.5.8, the Government shall advise the contractor, in writing, of the reasons for such failure or disapproval and the Government and the contractor shall meet to discuss the said development plan and the reasons for the said failure to approve or disapproval, and use their best efforts to agree on appropriate modifications thereto to meet the Government's concerns or objections. Thereafter, the contractor shall have the right to resubmit, within eighty-five (85) days of communication from the Government, the proposed development plan duly amended to meet the Government's concerns. Such right of resubmission of the proposed development plan shall be exercisable by the contractor only once. The Government will respond to the resubmitted plan within one hundred and fifteen (115) days. If no such plan is submitted to the Government within the above specified period, the contractor shall relinquish its right to develop such gas discovery and such discovery shall be excluded from the contract area.

21.5.10 In the event that the management committee or Government, as may be the case, approves the contractor's development plan for the development of such commercial discovery, with such modifications and amendments as the management committee or Government, as may be the case, may approve, the said gas discovery shall be promptly developed by the contractor in accordance with the approved plan which shall be the development plan.

21.5.11 The contractor will have a two (2) years period, from the date of approval of the development plan by the management committee or Government, to tie-up the market(s) for sale of non-associated natural gas.

21.5.12 In the event the contractor does not commence development of such discovery within ten (10) years from the date of the first discovery well, the contractor shall relinquish its right to develop such discovery and the area relating to such discovery shall be excluded from the contract area.

21.6 Valuation of natural gas:

21.6.1 The contractor shall endeavour to sell all natural gas produced and saved from the contract area at arm's length prices to the benefit of parties to the contract.

21.6.2 Notwithstanding the provision of Article 21.6.1, natural gas produced from the contract area shall be valued for the purposes of this contract as follows:

a. Gas which is used as per Article 21.2 or flared with the approval of the Government or re-injected or sold to the Government pursuant to Article 21.4.5 shall be ascribed a zero value;
b. Gas which is sold to the Government or any other Government nominee shall be valued on the terms and conditions actually obtained including pricing formula and delivery; and

(Explanation: However, it is clarified that this provision would apply only when the sale is made to the Government or Government nominee under the provisions of the contract.)

c. Gas which is sold or disposed of otherwise than in accordance with paragraph (a) or (b) shall be valued on the basis of competitive arm's length sales in the region for similar sales under similar conditions.

21.7 So as to ensure that the gas is valued at arm's length price or where arm's length price is impossible to arrive at the formula or basis on which the prices shall be determined pursuant to Article 21.6.2 (c) shall be approved by the Government prior to invitation of price bids or other price discovery steps by contractor for the sale of natural gas to consumers/ buyers, within sixty (60) business days from the receipt of the proposal or from the date of receipt of clarification/additional information, where asked for by the Government. For granting this approval, Government shall take into account amongst other relevant considerations, the domestic and international prices of comparable gas and the linkages with traded liquid fuels. The price of natural gas arrived at through the approved formula/ basis shall be applicable uniformly to all the consuming sectors indicated under Article 21.3.1. Where the contractor makes sale of gas at prices higher than approved formula/basis, then the higher price would be reckoned for the purposes of Government take.

4 Energy Efficiency

Introduction

Energy efficiency is achieved through R&D activities aimed at limiting the energy consumption per unit of activity. For example a century ago, a typical motor car consumed a litre of gasoline to travel six kilometres. But now, a typical family car consumes only 0.25/0.4 litres of gasoline to travel six kilometres. Similarly, machines are consuming lesser and lesser energy per unit of output. For example, energy consumption by refrigerators in India has reduced from over thousand KW per year to about 300 KW per year over a span of 30 years.

The Concept

Efficiency is derived from a more efficient utilisation of resources—in this case, the energy resources. The limited availability of energy resources necessitates less use of the energy for a given task. In other words, lesser amount of energy would produce a given amount of output than what it did previously. At the macro level, energy-efficiency measures would not hamper the rate of economic growth.

The demand for energy is derived from the activities it helps accomplish. Driving a given distance in an automobile, use of a motor for a specific period of time, running of a machine and accomplishing certain quantity of output all determine the quantity of energy demanded, the price determined and the value assigned to each source and form of energy.

An efficient energy strategy requires that the energy consumption should be rationalised and the losses should be reduced. Here is an example illustrating the rationalised demand for energy and the value of appliances using a given form of energy *viz.*, the electricity. There are two air-conditioned houses. The first house is well built and insulated with materials that restrict heat losses. The second house is poorly constructed and it permits the heat to escape. This shows that the first house is more energy efficient.

The above example shows that energy efficiency through conservation increases the economic wellbeing without compromising the level of output.

The Objectives

a) Avoiding the wastage and achieving higher economic efficiency.

b) Preserving energy resources for the use of the future generations.

c) Preventing environmental deterioration.

d) Mitigating the dependence on the foreign countries for energy resources: especially oil which is of greater contemporary relevance.

e) Enabling the use of energy resources for non-energy uses and

f) Postponing the depletion date of the non-renewable energy resources to allow more time for the development of technologies for more economic harnessing and use of renewable energy.

The Scheme

This chapter is organised into six sections. Section II describes the ways and means of achieving energy efficiency technically, structurally and behaviourally. Section III discusses the National Mission for Enhanced Energy Efficiency (NMEEE). Section IV discusses the policy, R&D and financing measures to fructify NMEEE. Section V discusses the National Mission on the Sustainable Habitat (NMSH) and the Management of Municipal Solid Waste (MSW) and the promotion of public transportation towards achieving energy efficiency and Section VI presents the concluding remarks.

Approaches to Energy Conservation

Energy efficiency can be improved by using the following measures.

a. Technical measures.

b. Structural measures.

c. Social and behavioural measures.

Technical Measures

Technical measures imply a change in the type of machine tools or implements that enable the use of energy. There are two basic kinds of technical approaches *viz.*, 'leak plugging' and 'machine switching.' Leak plugging eliminates the waste in existing technologies. Machine switching necessitates the replacement of existing devices with more efficient ones. For example, to insulate a house is to plug a leak and to replace an electrical resistance furnace with a heat pump is to switch machine. Similarly, to tune up a car is to plug a leak and to trade it for more fuel efficient model is to switch machine.

Technical measures take time to materialise to the desired extent because:

a) The fruition of R&D takes time.

b) Adequate financial and technological investment is required to bring about the change.

c) The economic cost of giving up the old technology should be less than the cost of the adoption of new technology.

The economic cost includes the interest rate, the depreciation, the operating cost and the energy saving cost *viz.*, the opportunity cost of substitution *viz.*, the residual value of old equipment.

Higher interest rates and the uncertainty in the capital markets tend to discourage the energy saving investment. Besides, the energy saving cost tends to decline faster only at the initial level of technological change and would increase thereafter.

Structural Measures and Social and Behavioural Measures

Structural measures warrant a change in the structure of the economy towards a low energy high growth path. A huge investment is required to effect this change. Initially, the marginal economic benefit is greater than the marginal economic cost. Overtime, both get equalised and later, the marginal economic cost exceeds the marginal economic benefit. The optimum level of investment is reached when the marginal economic cost equals the marginal economic benefit.

The economic benefit can be augmented with the increase in the price of energy: thus incentivising the energy saving investment. This shows that the cost of energy saving is less than the cost of energy provision without affecting the level of material enjoyment.

Social and behavioural measures bring about a change in the behaviour pattern of the people. The changed behaviour would require an alteration of lifestyle choices which in turn bring about a reduction in the consumption of energy in the enjoyment of material wellbeing.

Possibility of Technical Change

The scope of technical change is limited by reduced flexibility in effecting the change. The lifetime of capital equipment is so long that the residual cost of disposing the old capital can be a prohibitory factor. For example, an automobile and a house having a lifetime of about 20 and 50 years respectively would not be easily disposed off.

Energy conservation should either be made to its full potential at the beginning of the capital's lifetime or on an incremental basis based on the annual capital turnover (based on depreciation plus the additional costs). Fast growing economies can better conserve the energy achieving greater annual capital turnover.

In the case of the developing countries, 'leak plugging' is the best way to begin the energy conservation. The upward potential for energy efficiency in the developing countries is very high because they are still far behind the developed countries in terms of energy intensity and hence there is a lot to gain in terms of the 'thermodynamic efficiency' of the machinery.

Improving Technical Efficiency in the Petroleum Sector

Since many developing countries use petroleum, several practical measures are suggested by the Petroleum Conservation Research Association (PCRA) to improve efficiency:

a) Proper storage and handling of furnace oil to avoid spillage and leaks, to improve the heating and pumping system, to drain the water and sludge etc.

b) Ensure proper preheat temperature to maintain the correct viscosity at the burner tip and the provision of filters while preparing the furnace oil.

c) Advising on the choice of burners based on their flame profiles, operation of burners to achieve optimum excess air and studying the draft requirements and instructing operators on the finer aspects of draft control.

d) Advice on the correct instrumentation and control practices suited to the needs of the concerned industrial unit.

e) Evaluating the performance of the boilers and furnaces installed and recommending measures for economic loading and steps to reduce blowdown losses, radiation losses etc.

f) Having a steam distribution system, with proper layouts, correct size of pipelines, lagging, steam trapping and air venting, as well as returning the condensate to the process or to the hot wells.

g) Installing flash vessels to use the flash steam.

h) Improving the design of existing furnaces to ensure that the industrial furnaces especially forging units are built with chimneys, providing insulation bricks and ceramic fibres to keep down radiation losses.

i) Installation of waste heat recovery devices like recuperaters, regenerators and waste heat boilers and furnaces to recover waste heat from the hot flue gases.

j) Replacement of old, obsolete and inefficient equipment by modern, thermally efficient and less energy consuming units.

Of all the above recommendations, PCRA found in a 413-firm survey that waste heat recovery is the most productive method saving up to 10 per cent of fuel oils. An investment of Rs 30 crore would annually save about Rs 25 crore.

Improving the Technical Efficiency in the Agricultue and Transport Sector

In the field of agriculture, proper and informed choice of the pump sets and tractors by the farmers would result in substantial energy saving. In the transportation sector, better driving practices and regular maintenance can save six per cent of oil and natural gas. In developed countries, transport is said to be the major emitter of greenhouse gases (GHG). Government policies in several countries provided incentives to the developers of clean technologies related to fossil fuels leading to increased efficiency of vehicles and industrial and household equipment and appliances.

For example in the US, between 1965 and 1995, energy consumed for every dollar of constant GDP (1996 level) fell from 8,300 Btu (British thermal units) to 4,600 Btu. This implies that every four years, the fall was 500 Btu. However between 1995 and 1999, the fall was only 400 Btu despite tremendous improvements in the energy conservation technologies.

Between 1983 and 1999, over 40 per cent increase in the energy efficiency of automobiles occurred in US. Average mileage rose from 15.1 miles to 21.5 miles per gallon of gasoline used. There is a great scope for saving oil use in the US if the sports utility vehicles (SUVs) are targetted because SUVs comprise over 25 per cent of all the light vehicles in the US. Similarly in India, small cars dominate the four wheel market. Hence, by targetting the small car segment, substantial proportion of oil use can be saved.

To ensure an increase in energy efficiency, countries adopt time limits in the use of capital stock and enforce a cap on the energy emission. In the US, India, the European Union, coal-fired power plants should not function beyond 50 years and automobiles should not be used for over 15 years. Emission standards are determined by the Euro norms. Currently, 'Euro4' norms are the standard. Household and commercial sectors have good potential to conserve energy. The thermal efficiency of firewood-based stoves in rural household can be increased from 10 per cent to 20 per cent by adopting smoke efficient *chulhas*. Use of insulation and florescent lamps can conserve 25-40 per cent of the energy used.

Improving Technical Efficiency in the Electricity Sector: The Key to Achieving Optimum Utilisation of Energy

Power sector has a lot of scope to improve energy efficiency on the generation, transmission as well as on the distribution fronts.

On the Generation Front

Along with the transportation sector, power generation is another significant emitter of Green House Gasses. To curb emissions: especially through coal power plants, new technologies *viz.*, the integrated gasification combined cycle (IGCC) and pressurised fluidised bed consumption (PFBC) are being developed and applied in the industry. In India, IGCC is applicable in the existing coal power plants producing about 25 per cent of coal power *viz.*, over 28,000 MW. Apart from reducing CO_2 emission by about 0.69 million tonnes IGCC increases the 'thermal efficiency' of the coal power plant by about 20 per cent in turn increasing the 'gross calorific value' to 45-50 per cent. Thus, the 28,000 MW gets enhanced to 33,600 MW in the said power plants.

In the remaining coal power plants, the use of PFBC is encouraged. For one MW of power generated, PFBC technology reduces the carbon emission by 0.59 million tonnes.

When compared to the nuclear power which reduces 6 million tonnes of CO_2, IGCC and PFBC's contribution are minimal. However, nuclear power is said to have unquantifiable risks and hence opposed by environmentalists. Hence until the Wind and Solar power become cost competitive, IGCC and PFBC are the best possible options for energy conservation.

Government of India aims at ensuring quality and affordable energy to all the sections of the population in an environmentally acceptable manner on a sustainable basis. The Integrated Energy Policy of India encourages the use of all the forms and source of energy to sustain India's economic growth of 8 per cent or higher at least until the year 2031-32. The necessity of increasing the electricity supply by about four times the level of 2006 *viz.*, 1,44,000 MW and the necessity to increase the primary energy consumption by over three times is clearly envisaged in the 'Integrated Energy Policy.'

Relative scarcity of oil and potential environmental harm due to domestic coal are well recognised by the policy. Reduction of the share of coal in total electricity generation from 80 per cent to 60 per cent and proportionately increasing the share of wind power is a major challenge on account of cost. Hence, the use of existing 'clean coal' technologies *viz.*, the

IGCC and PFBC are being encouraged. Construction of new 'clean coal' power plants is encouraged near the coastal areas. Assured supply of coal is a major problem for these plants. Hence the Policy assures 20 per cent of coal to be made available through 'Electronic Auction'.

On the Transmission and Distribution Fronts

Over 40 per cent of rural household is denied access to electricity. Over 53 per cent of rural people *viz.*, the majority reported electricity is the major source of lighting. Since the Government of India aims at providing quality, reliable and adequate power supply to all the households by the year 2012, Government aims at enhancing the quality and accountability in the power sector. A strong and financially healthy power sector institution is essential to realise the policy objectives.

The problem of high aggregate technical and commercial (ATC) losses should be overcome to enable the power sector to mobilise the much needed funds to generate additional power. ATC deprives the power sector of the funds to the tune of 1.5 per cent of GDP with which over the period of five years India could produce an additional electricity of at least 100,000 MW using coal, 50,000 MW using wind power and over 20,000 MW of solar power respectively.

Between 2001 and 2006, the transmission losses were brought down to about 31 per cent from 32.6 per cent and the ATC loss was brought down from 38.5 per cent to 34.5 per cent respectively. In money terms, the ATC loss was brought down from over Rs 29,000 crore to Rs 19,000 crore respectively. In urban and high density areas, it is proposed that the ATC loss should be brought down to 15 per cent by 2010.

ATC loss is being overcome through infrastructural upgradation and poor transmission infrastructure and monitoring mechanism has to be done away with by introducing new technologies, net metering, geographical information system on power producers, transmitters and consumers. It is a major challenge. Hence to overcome this challenge, liberal and group captive regimes were introduced with wheeling charges being set for the transmission of power to the group captive consumers. To introduce wheeling charges, the wire carriage and pure content (power) are separately accounted and audited.

Interstate grid should be modernised to support high volumes of power transmission and to prevent load shedding. The grassroot infrastructure *viz.*, the transformers should be made up to date. Especially, the northern grid faces lot of problems and has poor connectivity to the other grids. Modernisation of the grid and the grassroot infrastructure enables the fixing of Availability Based Tariff (ABT) to the consumers in other grids.

ABT enables the introduction of open access to enable the power producers to auction the power to the highest bidders even in other grids. However, the incumbents *viz.*, public sector power companies complain that poor, low paying, bill evading and subsidised consumers would only remain with them and high paying consumers will switch to the private producers. To mitigate this problem, cross-subsidy charges, wheeling charges and backup charges should be properly set to realise the gains from the measures taken to mitigate the ATC losses.

Scope of Structural Measures

Structural Measures

The chief structural measure is the change in the mode of operation. For example, passengers can be encouraged to switch from personal to public transport for short distance and from road to rail transport for long-distance travel. Wherever applicable, passengers could also use the waterways.

Table 4.1 shows the gasoline required by mode of transport per mile of travel.

Table 4.1

Energy Costs of Transportation

Mode of Passenger Movement	
	Mile per US Gallon
Bus	1,090,125
Railroad	1,700,80
Automobile	4,250,32
Airplane	9,200,14

Source: See TERI Volume on Energy Conservation 2008.

Pipelines carry six times as much goods as the road and water and railway about five and four times respectively. However for short and moderate distances including for door delivery, roadways are the best. The goal of the public policy should be to execute pipeline, water and rail transport for long and moderate distances and roadways for short distance. As far as possible, personalised transport should give way to public transport.

Cost reduction must be the goal for a transportation system. Efficient use of raw material is important for cost reduction. For example, coal gives five times as much energy through electric traction as compared to the use of the coal powered steam engine. Hence, electrification of the railway lines should be complete. Although the capital cost is high, the operating cost is very low and routes with high traffic density are very economical for the railways.

Table 4.2

Mode of Freight Movement

	Energy Used Per Tonne Mile (Btu)	*Tonne Miles Per US Gallon*
Pipeline	450	300
Water	540	250
Railroad	680	200
Truck	2340	58
Airplane	37000	3.7

Source: See TERI Volume on Energy Conservation 2008.

Progressive electrification of the economy will help save the oil and natural gas which are relatively scarce. Moreover, the capacity factor of a diesel power generator is only 30 per cent while the coal power plant's capacity or plant load factor (PLF) is at least 65 per cent even in the worst case scenario in the country now. Economic and technical feasibility of electric vehicles will contribute to significant energy savings.

Social and behavioural changes like the optimum speed of driving, architectural changes allowing the inflow of natural light, switching off lights and devices when not in use, air conditioning at the optimal level of temperature etc., will go a long way in conserving the energy.

Promoting Energy Efficiency in India

Government of India is taking several important measures in promoting energy efficiency through various missions such as the energy efficiency, sustainable habitat, treatment of municipal solid waste (MSW) and the mitigation of GHGs by promoting public transport through the National Action Plan on Climate Change (NAPCC).

National Mission for Enhanced Energy Efficiency in Industry

Industry sector is the largest user of commercial energy in India. It accounts for 42 per cent of the country's total commercial energy use during 2004-05. Several of India's industries are highly energy intensive. Iron and steel, cement, fertiliser, refineries etc., are some of the examples.

The national greenhouse inventory records that the direct CO_2 emissions from industrial sources accounted for nearly 31 per cent of the total CO_2 emissions by India (data for base year 1994) (NATCOM I). The CO_2 emissions by the industry are of two broad categories *viz.*, process related and emissions due to fuel combustion in industries. Of the total estimated 250 million tonnes of direct CO_2 emissions from the industry in 1994, nearly 60 per cent were accounted for by energy use (NATCOM I).

GHG Mitigation Options in the Industry Sector

They are of three broad heads as given below:

a) Sector-specific technological options: Various GHG mitigation technology options in respect of the chlor-alkali, cement, aluminium, fertiliser, iron and steel, pulp and paper and textile sectors are currently being utilised and enhanced.

b) Cross-cutting technologies options: In the industrial sector, approximately 50 per cent of the energy use is accounted for by cross-cutting technologies. The estimated energy saving potential for a large number of plants is of the order of 5 per cent to 15 per cent under the current technological scenario.

c) Fuel-switch options: With the increasing availability of imported and domestic natural gas in India, industries have the viable option to switch over from coal to the use of natural gas. Fuel-switch to natural gas generally leads to an increase in the energy efficiency in use.

Industries with low temperature requirements (upto 100°C) (for example, textiles and pharmaceuticals) may switch over from fossil fuel to biomass-based technologies and may also use solar thermal systems for water heating.

Potential for Emissions Reduction

Although the efficiency of most large industrial sectors has been improving over time, and the specific energy consumption of many of the large plants compares well with the world's best, it is estimated that CO_2 emissions from fuel and electricity use in India's industry sector could be further reduced by about 605 million tonnes (approximately 16 per cent reduction from the 'business as usual' scenario) in the year 2031. This will, however, involve major incremental investment costs, large economic costs, besides technology transfer.

Co-Benefits

Reduced emission of air pollutants, solid waste and waste water are usually accompanied by the reduced use of fuel and material. Some technological options also lead to improvement in the quality of product.

Technology Transfer

Technology transfer is to be effected through inter-firm MoUs and inter-governmental MoUs along with the MoUs between the industries and between the industries and the government.

Financing

The switch-over to the energy-efficient technologies in the industry sector generally involve significant incremental investment, and in many cases, economic costs. Multilateral funding arrangements needed to be worked out to enable financial availability. Bundling or programmatic clean development mechanisms (CDMs) would help formulate appropriate models involving subsidies, interest rate rebates and preferential treatment of small and medium industries.

Capacity-Building Needs

Cooperation between the government and industry is imperative to enhance awareness of energy-efficient options and to upgrade relevant technical knowledge. The financial sector needs capacity building in the appraisal of specific energy-efficiency improvement investments.

Policy and Regulatory Options

Energy Audit

Under the Energy Conservation Act of 2001, nine energy intensive industries *viz.*, thermal power stations, fertiliser, cement, iron and steel, chlor-alkali, aluminium, railways, textile and pulp and paper, are required to employ a certified energy manager, conduct periodic energy audits and adhere to specific energy-consumption norms that may be prescribed.

Currently, almost every industrial sector is characterised by a bandwidth of energy efficiencies. Several of them are at the global frontier. However, some other sectors perform rather poorly. The energy-efficiency bandwidth is divided into four bands. The target bandwidth is highest for the least energy-efficient industry and least for highest energy-efficient industry. Within each industry group, targets would have to be achieved within a period of 3 to 5 years.

Incentive Mechanism

In the industrial units dependent on Government subsidy—such as the fertilisers, energy efficiency can be improved with proper Government incentives. Subsidies should be restructured and retargetted to benefit the firms genuinely undertaking the best possible energy-efficient measures.

Technology Development Programmes

To promote technology upgradation in the small and medium enterprise (SME) sector, it would be essential to evolve sector-specific integrated programmes for technology development. This would require external support for significantly longer durations to overcome various technological barriers and promote energy efficiencies at the unit level. The information or the knowledge gap is more pronounced in the small industries. They require 'hand-holding' towards the installation of energy-efficient technologies as well as towards ensuring their optimum performance through best operating practices.

Improving Solvency

Most of the energy-efficient equipments require higher upfront investment. Policy level support is required for making the equipment cost-effective. An accelerated depreciation of up to 80 per cent in the first year and the reduced rate of value added tax (VAT) would promote their rapid deployment.

New Initiatives to Enhance Energy Efficiency

To further enhance energy efficiency, four new initiatives may be considered. These are:

a) Mandated specific energy consumption decreases in large energy consuming industries and facilities that have been notified as designated consumers under the Energy Conservation Act, and provide a framework to certify energy savings in excess of the mandated savings. The certified excess savings may be sold to companies to help the defaulters meet their mandated compliance requirements, or banked for the next cycle of energy savings requirements.

b) Tax incentives for promotion of energy efficiency, including differential taxation on appliances that have been certified as energy efficient through energy labelling programme.

c) Creation of energy-efficiency financing platforms for enabling public-private-partnerships (PPPs) to capture energy savings through demand-side management programmes in the municipal, buildings and agricultural sectors.

d) Fiscal incentives: Tax rebates should be given to consumers for the purchase of appliances that are energy efficient and which qualify for star ratings.

Delivery Options

The key delivery options for energy efficiency in industry include:

a) Projects, including retrofits, by the corporate sector, with institutional finance.

b) Activities related to cluster development, particularly in small and medium industries.

c) Promotion of Energy Service Companies (ESCOs) for providing energy-efficiency solutions across industry sectors.

The Energy Efficiency Financing Platform (EEFP) initiated by the Bureau of Energy Efficiency (BEE), in conjunction with a robust ESCO industry could provide the necessary impetus to energy efficiency. Each delivery mode should encompass carbon finance through the CDM mechanism.

Energy Efficiency through the National Mission on Sustainable Habitat

The Mission comprises three components *viz.*:

a. promoting energy efficiency in the residential and commercial sector,

b. management of municipal solid waste and

c. promotion of urban public transport.

Promoting Energy Efficiency in the Commercial and Residential Sector

The residential sector accounts for around 13.3 per cent of total commercial energy use in India. While several households, especially in the rural areas, continue to use biomass for cooking in traditional cook stoves, which in turn leads to high levels of indoor air pollution and poses a major health risk especially to women and children due to health reasons; and due to the increase in the Government initiatives in the form of subsidies, the use of modern fuels such as liquefied petroleum gas (LPG) and kerosene has been increasing rapidly. During 1990-2003, consumption of LPG increased at an annual rate of 11.26 per cent, while electricity use increased at 8.25 per cent annually in the residential sector.

Electricity consumption in the residential sector is primarily for lighting, space conditioning, refrigeration, and for the use of other appliances. According to a study on energy consumption in the residential sector in the city of Delhi, while lighting accounted for around 8 per cent-14 per cent of total electricity consumption, space conditioning accounted for nearly 52 per cent, and refrigerators accounted for around 28 per cent (in the summer months). Hence, energy saving measures concerning space conditioning (heating and cooling), refrigeration and lighting contribute significantly to sustainable residential energy use.

The commercial sector comprises various institutional establishments such as banks, hotels, restaurants, shopping complexes, offices and public buildings. Electricity consumption has increased at the rate of 7.4 per cent annually between 1990-2003 in the commercial sector. It is estimated that on average, around 60 per cent of the total electricity is consumed for lighting, 32 per cent for space conditioning and 8 per cent for refrigeration in a typical commercial building in India. However, the end-use consumption varies significantly with space conditioning needs.

A fully air-conditioned office building could use about 60 per cent of the electricity for air conditioning and 20 per cent for lighting in a non-air-conditioned building, the consumption patterns would differ significantly. Energy use in residential and commercial buildings also varies significantly across income groups, building construction typology, climate, and several other factors. A significant scope exists to reduce energy use keeping the level of comforts same or in some cases on a higher level.

Although the saving potential of each option may vary with typology, climate, space conditioning needs, and the initial base design proposed by the client/designer, on an average it is estimated that the implementation of energy-efficient options would help in achieving around 30 per cent electricity savings in new residential buildings and 40 per cent electricity savings in new commercial buildings. In case of existing buildings, the energy saving potential for residential buildings is estimated to be around 20 per cent, and that for commercial buildings around 30 per cent.

Implementing carbon mitigation options in buildings result in a wide range of co-benefits. They include: improved energy security and system reliability along with the increased availability for poor households and small enterprises and improved air quality leading to improved health benefits on account of reduced level of pollution.

Costs and Financing

The incremental cost of implementation of energy-efficient measures is estimated to vary between 3 per cent–5 per cent for residential buildings and 10 per cent-15 per cent for commercial buildings on a case-by-case basis. Usage pattern determines the economic lifetime of such equipment. However, private owners tend to seek a shorter payback period *vis-à-vis* the commercial property owners. Appropriate financing mechanisms help mitigate the repayment burden arising out of increased upfront cost of investment.

Energy-efficient lighting and space-conditioning technologies should be integrated into housing finance schemes of financial institutions. Appliance financing schemes should incentivise the purchase of energy-efficient equipment. Utility-based programmes should be put in place to pay for the higher upfront capital costs of lighting systems in the utility bills. Carbon market financing would enable access to the technologies where there are higher investment costs, or higher economic costs of the required energy service, or both. This kind of scheme would help solve the 'split incentive' problem.

Split incentive problem arises because the persons who incur the additional investment costs are different from the persons who might realise the energy savings.

Research & Development (R&D)

The R&D needs for the residential and commercial sectors is mainly related to energy-efficient technologies. It needs to focus on the development of energy-efficient products for the following applications:

a. Energy-efficient buildings and building components,
b. Development of energy efficient windows,
c. Development of low-cost insulation material,
d. Development of simulation software to predict the energy used in buildings,

e. Energy-efficient appliances,
f. Development of energy-efficient ceiling fans,
g. Development of very low-energy consuming circuits for standby power and
h. Development of low-cost light-emitting diode (LED)-based lamps for space lighting.

The Scientific Advisory Committee of the Cabinet (SAC-C) has recommended the launch of a national networked initiative for R&D on the development of the next generation of LEDs, particularly white LEDs.

Technology Transfer and Capacity Building

The energy-efficient lighting and space conditioning technologies developed internationally are generally superior as compared to those available within the country. There is therefore a need for technology transfer from the developed countries. However, adopting these internationally developed technologies is associated with payment of additional costs due to the Intellectual Property Rights (IPR) component associated with these technologies.

Policy support mechanism should ensure that the burden of the additional costs should not impose an additional burden on the consumers. For example, solar evacuated tubular panel technology is available internationally for solar water heating systems, but needs to be transferred for diffusion in the Indian market.

Lack of awareness of energy-saving options and potential among architects, engineers, interior designers and professionals in the building industry including plumbers and electricians is a major barrier to the construction of low-energy buildings. Realising the potential of energy saving requires an integrated design process involving all the stakeholders, with full consideration of opportunities for passively reducing building energy demands.

Builders and developers need to be trained and made aware of the options to save energy in new constructions. There is a need to create comprehensive integrated programmes at universities and other professional establishments to impart such training for designing and constructing low-energy buildings.

Policy and Regulatory Enhancements

A diverse portfolio of policy instruments would be required to address the barriers to efficient energy use in the residential and commercial sectors. Firstly, there is a need to continuously update appliance energy norms and building energy codes and labelling. Secondly, a gradual shift should be made towards a rational energy pricing based on long-term average economic cost. Thirdly, fiscal benefits should be provided for improvement in energy efficiency in residential and commercial areas.

The Energy Conservation Building Code (ECBC) was developed after the adoption of the Energy Conservation Act (2001). It aims to reduce the baseline energy consumption by supporting adoption and implementation of efficiency measures and saving in GHG emissions, besides providing other benefits. ECBC intervention has encouraged design innovation in the building envelope and system design and specification. Such measures have resulted in 50 per cent energy savings (in ECBC compliant buildings) *vis-à-vis* in the conventional constructions.

Energy savings should be mandated through policy mechanism and using the ECBC. The costs of compact fluorescent lamp (CFL) and efficient tube light (T5) should be brought down by ensuring the increased demand and consequently harnessing positive economies of scale. Appropriate materials and equipment should be available in adequate quantity to meet the ECBC requirement. The code itself is new in India and the products such as, insulation, efficient glass, efficient heating, ventilation and air-conditioning (HVAC) systems and services etc., are neither readily available nor 'competitively priced, insulations, chillers and other products are manufactured by few companies.

In addition to the above, the Ministry of Environment and Forests (MoEF) has developed a manual on norms and standards for environmental clearance for large construction projects after wide consultation with experts from different disciplines. The manual would serve as a technical guideline to assist the project proponents, stakeholders and the consultants in carrying out the environmental impact assessments of projects and obtaining environmental clearances.

Both the Expert Appraisal Committee (EAC) of the MoEF at the national and State Expert Appraisal Committees (SEACs) at the state and union territory level appraise and grade all new construction projects requiring environmental clearances on the basis of the manual. The state pollution control boards are required to verify the adherence to the Environmental Management Plan and the observance of the criteria of gradation by the project proponents.

Successful implementation of performance-based codes requires education and training of building officials and inspectors and demonstration projects. Performance-based codes can better help keep compliance costs low and may serve as incentives for innovation as compared to the imposition of technology.

Delivery Options

The Bachat Lamp Yojana (BLY) model needs to be pursued to promote energy efficient and high quality CFLs which should replace the incandescent bulbs in households. This can lead to a reduction of 10,000 MW of electricity use or about 6.5 per cent of India's current electricity production

Financing mechanisms should meet the incremental investment cost as well as the incremental economic cost. ESCOs need to be promoted as vehicles to deliver energy-efficiency improvements to overcome the 'split incentives' problem, and to facilitate access to carbon finance through bundled CDM projects.

Management of Municipal Solid Waste (MSW)

Municipal solid waste (MSW) generation reflects the income levels of a country/city along with the lifestyle choices. Recycling of materials is an important option for reducing adverse environmental impact. India has a significantly higher rate of MSW recycling than developed countries. GHG emissions from MSW in India are much below the developed countries in $ 1000 at purchasing power parity (PPP). MSW generation in Indian cities is estimated to have increased from 6 million tonnes in 1947 to 48 million tonnes in 1997 and to 69 million tonnes in 2006 (Central Pollution Control Board, 2000, TERI, 2001).[1]

---

1. Central Pollution Control Board Municipal Solid Waste Management and Handling Rules 2001 and TERI Energy Data Year Book 2001.

In addition, Indian consumption of plastics is around four million tonnes per annum (MTPA). About 60 per cent of this comprises polyolefins, which are primarily used as packaging material. About 2.0 MTPA of total consumption becomes part of the MSW of which around 70 per cent is recycled, mostly by the informal sector. The growth in consumption of plastics between 1991-2001 was around 14 per cent (Indian Centre for Plastics in the Environment and Central Institute of Plastic Engineering Technology, 2003).[2]

The quantity of plastic waste reaching disposal sites is fairly low (0.62 per cent on a dry weight basis). The high rate of recycling/reuse has not resulted in the better management of thin plastic bags due to low collection efficiency necessitating the strengthening of this segment.

Table 4.3

Characteristics of MSW in 59 Cities

Parameter	*Unit Range*
Compostable (%)	30 - 73
Recyclable (plastics, paper, metal, glass etc.) (%)	10 - 37
Moisture (%)	17 - 65
Carbon/nitrogen (C/N) ratio	14 - 53
HCV Kcal/kg	520 - 3766

Source: CPCB, 2005.

There is an increasing trend of the percentage of recyclables and a decreasing trend of the percentage of biodegradable matter in the waste stream.

Policies and Regulations

The 74th Constitutional Amendment (1992) transferred the responsibility for collection, treatment and disposal of MSW from state governments to the urban local bodies (ULBs). The outbreak of plague at Surat in 1994 increased the policy focus on the importance of proper

2. Statistics section of the Indian Centre for Plastics in the Environment (Accessed approximately in December 2010) and Central Institute of Plastic Engineering Technology (CIPET) Data Book on Plastics 2003.

systems for MSW treatment in the ULBs. A public interest litigation (PIL) (WP No. 888/1996) resulted in the promulgation of the MSW Rules, 2000.

Table 4.4

Change in Waste Composition in Selected Cities

City	*Compostables (%)*	*Recyclables (%)*	*1982-1990–2005*	*1982-1990–2005*
Lucknow	60.31	47.41	6.72	15.53
Kolkata	46.58	50.56	2.58	11.48
Kanpur	53.34	47.52	2.57	11.93
Mumbai	59.37	62.44	3.85	16.66
Delhi	57.71	54.42	8.24	15.52
Chennai	56.24	41.34	6.60	16.34
Bangalore	75.00	51.84	2.70	22.43

Source: Planning Commission, 1982-1990; CPCB, 2005.

According to them, MSW service from generation to disposal was mandated and local governments were made responsible for compliance.

Not withstanding the improvement in the collection and transportation of MSW, major gaps exist in treatment and disposal. The compliance is poor (<5%), due to negligence and poor capacity. Efforts at composting and generating energy from waste are less successful due to systemic, technological and pricing related issues. Variability in the quality of MSW, insufficient segregation of MSW, local opposition to site selection and persistent open dumping hinder the successful treatment of MSW.

Since the commercial viability of MSW treatment is not assured, ULBs should incur expenditure to cover the revenue gap while at the same time focussing on cost-effectiveness. The MSW Rules under the Environment Protection Act are focussed more on specific treatment options, including the chain of collection, transport and disposal. Such restricted focus tends to prevent innovation in systems and procedures, as well as the update on new technologies and techniques.

The targetted performance or outcome norms should become the focus irrespective of systems, structures, technologies etc. This would benchmark the monitoring indicators leading to proper enforcement and enable innovations in systems, procedures and technologies. To enable quality and cost-effective services for smaller habitations, the sharing of infrastructure and facilities is being contemplated. The establishment of integrated facilities with the differentiated management of the components, public-private partnerships to achieve cost-effectiveness and better accountability of ULBS is important for the viability of sustained MSW treatment.

The National Environment Policy of 2006, provides for:

a) Removal of barriers and promulgation of regulations for the beneficial utilisation of non-hazardous materials.
b) Implementation of viable PPPs for operation of hazardous and non-hazardous waste disposal facilities on payment of user fees taking local communities into confidence.
c) Survey and preparation of national inventory of toxic and hazardous waste sites and online monitoring of their movement.
d) Legally recognising and strengthening the informal sector systems of collection and recycling and enhancing their access to finance and technology leading to their upscaling and the realisation of full potential.

Areas of R&D Needs

a) Biomethanation technology for waste to energy including its decentralised application for segregated waste streams like vegetable market waste, slaughterhouse waste and dairy waste.
b) Development of indigeneous gas engines for waste to energy applications to reduce the overall cost of the package.
c) Upgrading plastic waste recycling technologies to reduce occupational and environmental hazards.
d) Recycling technologies for construction and demolition of wastes and waste streams.

Promotion of Urban Public Transport

Economic growth gives rise to an increase in the growth rate of passenger and freight transport. The number of registered motor vehicles in India has increased from 21.4 million in 1991 to 72.7 million in 2004 at a CAGR of 9.9 per cent, the two wheeler segment grew most rapidly.

Road-based transportation is the main source of GHG emissions. Various studies have estimated that policy and technological measures can lead to significant energy and emission savings in the transport sector. Estimates of the Planning Commission indicate an energy-saving potential of 115 Mtoe in the year 2031/32 by increasing the share of railways and improving the efficiencies of different modes of transport (Planning Commission, 2006). Similarly, TERI estimates indicate an energy saving of 144 Mtoe in 2031 by improving the efficiencies across the modes and by using the public transportation, including the railways. If biodiesel replaces the fossil fuels substantially, the CO_2 emissions are estimated to decrease by 433 million tonnes in 2031.

Transport Options

Mass transport options including buses, railways and mass rapid transit systems, etc., are the principal option for reducing energy use in the urban transport sector, and mitigating associated GHG emissions and air pollution. The use of CNG has helped reduce air pollution by replacing the use of diesel in some cities. Ethanol blending of gasoline upto five per cent is required in nine states, and is expected that this limit would be increased to 10 per cent thus increasing the penetration of biofuels.

R&D has to be carried out on the combustion characteristics of motor engines for blending of higher content of ethanol in petrol. Biodiesel production from *jatropha curcas* and *pongamia glabra* shrubs is increasing in quantum. R&D should increase the oil content of biodiesel crops and should help enable the identification, inventorisation and documentation of superior genotypes whose net GHG emission should not be positive.

Hydrogen has the potential to replace fossil fuels in the future. In recent years, significant progress has been reported by several countries for overcoming problems in the storage and production of hydrogen. In India, a National Hydrogen Energy Road Map (NHERM) has been prepared. The actual experiments are limited to prototypes whose full scale operationalisation is expected only after several decades.

Costs and Financing

Significant cost of infrastructure is a barrier to the implementation of mass rapid transit system (MRTS). This apart, other public priorities have to be sacrificed by the Government. However, the reduced GHGs can reduce the public health expenditure in the long-run possibly offsetting the costs of MRTS. The cost of alternative fuels is still much higher.

Co-Benefits

Increased public health benefits, reduced GHG emissions, reduced noice levels and the reduction in accidents on account of reduced pollution related fog and smog are the significant co-benefits. Besides, the reduced public health and road traffic related expenditure help improvement in the quality and quantity of public transport.

Recycling of the vehicles will reduce the use of steel, aluminium, rubber and plastics by 1.5, 0.18, 0.075 and 0.075 million tonnes, respectively, significantly contributing to conservation of raw materials and prevention of mining related environmental deterioration. The following actions would improve the co-benefits from the transport sector:

a) Promoting the use of coastal shipping and inland waterways, and encouraging the use of railways for long-distance.
b) Encouraging energy R&D in the Indian Railways.
c) Introducing appropriate transport pricing measures to influence the purchase and use of fuel-efficient vehicles and vehicles using various forms of fuels.
d) Tightening of regulatory standards such as the enforcement of fuel-economy standards for automobile manufacturers.
e) Establishing mechanisms to promote investments in the development of high capacity public transport systems.
f) Usage and abandonment of old vehicles should be made illegal and legislation should fix the responsibility of handing over the end-of-life vehicle to collection centres upon the last registered owner.
g) Setting up of a demonstration unit to undertake the recycling of vehicles, especially two wheelers using new technology.
h) Setting up a Combustion Research Institute to facilitate R&D in advanced engine design.
i) Providing tax benefits and investment support for the recovery of materials from scrap vehicles.

Promotion of energy efficiency takes place through a combined strategy of mitigating fuel use, promotion of fuel switching and through the development of multipurpose technologies. The current pattern of energy use and the possibilities and potential for energy conservation can be understood through energy auditing. There is great scope for energy auditors in India. The necessity for energy audit at least once in five years is imperative for almost all the industries and not only for the nine designated industries because it is in the interest of every individual to save energy.

Even at the household level, energy auditor can help realise the energy saving of over 20 per cent or at least an electricity bill reduction of about Rs 2,000-3,000 for middle class and even up to Rs 20,000 for richer households by advising on the technical change in the form of 'leak plugging' or occasional 'machine switching' and even on a change in the behavioural pattern.

A typical income tax auditor collects Rs 500 as fee from a typical middle class household. An energy auditor could collect about Rs 200 and serve about 4,000 households a year. Thus, there is a scope to employ at least 10,000 household energy auditors at a time. Either they can be employed as private consultants or by the shops that sell electric and electronic appliances to the consumers.

Energy efficiency can be improved through R&D activities in the areas of combustion, biomethanation, waste to energy application, plastic recycling, etc., increasing the yield of oil from oil seeds like the jatropha etc. Indian automobile sector is poised to assimilate the automobiles with Euro4 norms thus enhancing fuel efficiency by at least 30 per cent for four and two wheel vehicles and above.

Concluding Remarks

The world has made great strides in enhancing energy efficiency. In the past 100 years, the world has saved 60-75 per cent of gasoline in running a motor car. Similarly, the energy intensity of the economies has reduced by more than 2/3rd. Even in India, the energy efficiency has reduced tremendously. There is a great scope for reducing wasteful energy consumption and promoting the energy efficiency. Even the municipal solid waste can be converted to energy through the biomass power plants.

R&D initiatives in India have made some strides in promoting energy efficiency in domestic, commercial and industrial establishments. In the household, modern construction methods can enhance the energy efficiency by up to 40 per cent through better insulation and through the prevention of the leakage. In the commercial and in the industrial establishments, up to 30 per cent and up to 20 per cent efficiency can be achieved.

The usage of public transport and particularly the rail transport is heavily recommended because, the per passenger fuel saving is up to 40 per cent. Moreover, the direct use of coal and the diesel wastes over 60 per cent and over 50 per cent of energy. Hence, it is recommended that all the transportation should be run using the electricity. Even the industrial establishments should use oil to the best possible extent.

However, oil has a role to play for another 40-60 years and natural gas 60-100 years respectively in the area of transportation and industrial use depending on the estimates. The pressing concerns faced by the world are the sustained increase in the production and supply of oil and natural gas.

5 Energy Trading

Introduction

Trading in energy resources is an age old phenomenon. From the days of the kings whose urban settlements and armies needed to purchase the firewood and animal fodder to the modern age in most of the countries in the world need to buy oil and natural gas to circumvent the fuel inadequacies, energy has become the most sought after commodity. Without the energy, the household, commercial and industrial activities will come to a grinding halt.

The modern lifestyle is so dependent on oil and natural gas that the huge population of about seven billion cannot be fed without the type of economic activity driven by these fuels. Even if the world has to decide upon the firewood as an energy choice in the event of severe shortage of oil and natural gas, it is not feasible.

Out of the nearly seven billion people in the world, nearly two billion depend on firewood and they face shortage even for the rudimentary activities like cooking and warming themselves and their family members. Apart from the physical shortage of the firewood, the natural constraints in the form of low thermal efficiency restrict the scope of its usage as a fuel for economic activity.

Besides, the collection of firewood consumes a significant part of the working time of the day: especially for the women. Firewood is also the cause of respiratory illness for the women, children and the elderly. In the modern age where the Governments are responsible for public health and literacy, prevention of respiratory illnesses and promotion of schooling and women's economic activity become paramount public policy concerns.

Promotion of the usage of natural gas at the household level saves physical labour in the collection of firewood, provides time for women to participate in the economically beneficial activities and also better care for their children including ensuring their education. Besides, it drives forward the nation's economic activity. The firewood is short in supply due to deforestation. At the beginning of the 19th century, the world had a population of only six hundred million and now it over ten-fold.

Naturally, firewood-based economy can only support the subsistence of six hundred million people and superior energy resources are required to support a higher population base. Coal could only support the economy driven by about 1.5 billion people and oil by about four billion people. Natural gas is necessary to support an economy driven by over six billion people. To support an economy driven by over 10 billion people on a sustainable basis, economically and abundantly available renewable energy including the nuclear energy is imperative.

The world's population could stabilise at a level of about 10 billion or 13 billion people depending on 'mother earth's' feeding capacity and people's religious sentiments. For the next 40 years at least, a full-fledged transition to renewable energy is not possible due to technological and cost constraints. Dependence on the optimally allocated oil and natural gas is imperative for the economies to continue to grow and eliminate poverty. It is in this backdrop, this chapter on 'energy trading' is written. The next chapter exclusively focusses on 'power trading.'

Starting with Introduction, Section II focusses on the trading of oil, maintenance of strategic reserves, the role of OPEC in price stabilisation and the future of oil. Section III focusses on the trading and pricing of natural gas, the increasing dependence of the world on natural gas, the impending role of the Middle Eastern region as the price stabiliser and the possible surprises in store in the form of new and huge discoveries of natural gas. Section IV presents concluding remarks.

Oil Trading

The price of oil is traditionally determined by the matching or un-matching of the supply with demand. Excess supply leads to competition among the suppliers leading to lower prices. In turn, excess demand leads to competition among the consumers leading to higher prices. Along with the current supply/demand balance, expectations of the buyers and sellers concerning the future behaviour of prices tend to determine the present price level.

If buyers think that supply would fall short of demand in a few weeks or months leading to an increase in the price level, they would wish to stockpile certain quantity of oil presently and would even pay a premium today. If the supply is expected to increase soon, either the consumers would delay their purchases or would demand a price discount.

Along with the forces of demand and supply, price level tend to be determined by the modern screen trading mechanisms and online news services which might or might not reflect the reality. Nevertheless, these modern sources are so potent that they constantly play upon the minds of the consumers rapidly changing their trajectory of expectation.

Demand for Oil

The demand for oil is closely linked to the growth trajectory of the world's economy. Recently, oil demand has been growing fastest in the newly industrialising countries of the Asia-Pacific Rim that are experiencing rapid economic growth.

The OECD countries still experience more than half of the global demand for oil and the countries are experiencing more competition and least Government controls on the oil trade. Economic growth apart, the demand for oil is highly seasonal. Demand for heating oils *viz.*, the kerosene, gas oil and residual fuel oil peaks in the winter, while the demand for transport fuels such as gasoline and diesel peaks in the summer. Products such as bitumen (used for road building) also display a clear seasonal pattern influencing the oil price behaviour at certain times of the year.

Transportation tends to be the dominant influencing factor of the demand for oil. Yet, there is a three to four million barrels of daily variation of demand that influence the price level. The impact of oil price changes occur in the long term *viz.*, in over five years because it takes time to change the boilers or automobiles.

However, electricity generation is affected almost immediately and the use of alternate fuels *viz.*, the coal, natural gas etc., can be augmented in the short run. Tax levels tend to determine the level of oil consumption in countries with similar per capita incomes. For example, the US citizens consume nearly twice as much oil as compared to the citizens of the UK, France etc., on account of lower tax levels.

Demand Determination

Various factors determine the level of demand for oil:

1. *Economic factors*
   a) Levels of production & consumption levels,
   b) Availability of storage space and its cost,
   c) Cost of insurance,
   d) Rate of interest,
   e) Cost of transportation,
   f) Rate of currency exchange and
   g) The behaviour of other related markets.
2. *Geographical factors*
   a) Availability of pipelines and shipping,
   b) Localised weather report,
   c) Seasonality and
   d) Locality/terrain.
3. *Comparative factors*

Availability of alternative fuels enables the shift of preferences. This shift of preferences is determined by:

a) Transaction costs,
b) Alternative commodities,
c) Product yields and
d) Relative quality.

4. *Political factors*

   They include:

   a) Cartels/trade agreements,

   b) Global disturbances like war, embargoes etc.,

   c) Domestic taxation and trade tariffs,

   d) Regulations legislations and

   e) Environmental pressure.

Supply of Oil

Supply of oil is determined by the level of sunken investments in the oilfields and technical constraints on increasing the oil output. During the 1980s and at times during the 1990s and in the first decade of the 21st century, lower operating cost meant that there was great attempt to maximise profit during the short term: especially by the UK/Dutch consortium in the North Sea and by the Saudi Arabia.

However, most of the OPEC countries agreed to cooperate in the attempt to stabilise the prices in order to obtain fair economic returns. The fair economic returns on the oilfields have become difficult due to the expansion in the production in the non-OPEC countries which focus more on the technological advancement and cost countries.

The Sakhalin oilfields of Russia and the emerging oil producing countries of Africa are the examples. The seasonal variation in the demand pattern makes it difficult to enforce compliance among the OPEC countries with regard to fixity in production. The fixation of quota cannot be strictly implemented to let the market behave freely. Political pressures in the oil consuming countries and the need of the OPEC countries to diversify their economies with the help of OECD and other major countries like India to sustain their living standards and development makes the enforcement of quota very difficult.

The Determination of Oil Stocks

Oil stocks are determined by the expected supplies and by the geopolitical security concerns. The 1980s and a good part of 1990s were characterised by higher supplies on account of the discovery of the North Sea oilfields and the expectation of piece. The Gulf wars of 1991 and 2003, the Somalia conflicts from the 1990s until now, the Afghan wars and the increase in Islamic fundamentalism prompted the countries to increase their strategic stock piles. Oil stocks in US and other OECD countries substantially declined in the 1980s from about 90 days to about 55 days and are increasing to their former levels. Even India wishes to increase its oil stock piles to 90 days gradually. The US plans to have a 180-day strategic oil stock pile.

World Oil Markets

The world oil industry is largely concentrated. Top 10 companies control 68 per cent of the world's proven oil reserves. Nine of the top 10 companies are state-owned. OPEC countries account for about 70 per cent of world's oil reserves and accounted for one per cent of world's oil supply in 2005.

From the 1960s until now, the OPEC countries merely played the role of price stabilisation. Their future role of price determination through market domination is soon emerging. Canada holds the second largest oil reserves in the world, with over 178 billion barrels. Over the next decade, Canada will become a leading oil producer. The oil sands production in Canada would triple.

Oil Consumers

Consumers of oil continue to be the industrialised *viz.*, the OECD countries. The United States, Europe and Japan together consume about half of the world's annual oil output. Emerging markets like the China and India are experiencing faster economic growth rate. Transportation accounts for about 2/3rd of world and half of US oil consumption.

Table 5.1

Major Oil Consuming Countries in 2004

United States of America	(20.7 million barrels/day)
China	(6.5 million barrels/day)
Japan	(5.4 million barrels/day)
Germany	(2.6 million barrels/day)
Russia	(2.6 million barrels/day)
India	(2.3 million barrels/day)
Canada	(2.3 million barrels/day)
Brazil	(2.2 million barrels/day)
South Korea	(2.1 million barrels/day)
France	(2.0 million barrels/day)
Mexico	(2.0 million barrels/day)

Source: TERI Volume on Energy Trading 2008.

Oil Trading

Table 5.2

Major Oil Exporters in 2004

Saudi Arabia*	(8.73 million barrels/day)
Russia	(6.67 million barrels/day)
Norway	(2.91 million barrels/day)
Iran*	(2.55 million barrels/day)
Venezuela	(2.36 million barrels/day)
United Arab Emirates*	(2.33 million barrels/day)
Kuwait*	(2.20 million barrels/day)
Nigeria	(2.19 million barrels/day)
Mexico	(1.80 million barrels/day)

Source: TERI Volume on Energy Trading 2008.

Table 5.3

Major Oil Importers in 2004

United States of America	(12.1 million barrels/day)
Japan	(5.3 million barrels/day)
China	(2.9 million barrels/day)
Germany	(2.4 million barrels/day)
South Korea	(2.2 million barrels/day)
France	(1.9 million barrels/day)
Italy	(1.7 million barrels/day)
Spain	(1.6 million barrels/day)
India	(1.5 million barrels/day)
Taiwan	(1.0 million barrels/day)

Source: TERI Volume on Energy Trading 2008.

India usually imports oil from Venezuela, Nigeria, Sudan, Iran, Kuwait and Russia.

World Oil Price Trend from the Post-World War II Period

Oil price trends are derived from the world oil series that derives the prices at the wellhead which is consistent with the refineries acquisition price. Comparing with the 2006 dollars, the average oil prices between 1948 and 1957 ranged between $14 and $17 and stabilised at $17 between 1957 and 1970. Reduced level of inflation and the weakening of the US dollar brought the real price down to $14 in 1971-72.

OPEC was formed in 1960 with five founding members: Iran, Iraq, Kuwait, Saudi Arabia and Venezuela. Two of the representatives studied the Texas Railroad Commission's methods of influencing price through production limitation. By the end of 1971, six other nations joined OPEC *viz.*, Qatar, Indonesia, Libya, United Arab Emirates, Algeria and Nigeria.

Between 1948 and 1972, the oil exporting countries experienced a decline in the purchasing power of a barrel of oil by about 40 per cent. By 1972, the US oil producers were lacking in spare capacity and hence could no longer influence the price level. This coupled with the US support for Israel in the 1973 war made the OPEC countries realise the extent of their power in controlling oil prices which led to a three-fold increase in the price of oil. The Islamic Revolution in Iran in 1978-79 and the Iran-Iraq war from 1980 cut the oil production by half more than doubling the price of oil from $14 in 1978 to $35 in 1980. Even in 2005, Iran's oil production was only 2/3rd of the level achieved before the Islamic Revolution.

The US policy of price control over the domestic oil production led to the subsidisation of domestic consumers consequently delaying the energy conservation measures. The relatively lower oil prices paid by the US consumers between 1973-1979 led to increased oil consumption in turn leading to a rapid increase in the world oil price in 1979.

The Relationship between the Oil Price and Economic Performance

The 1979 oil price increase has brought about a more permanent and lasting results in terms of increased energy efficiency in homes, in automobiles and in the industrial processes. Oil producers would no longer rely on price reduction as the strategy to increase the demand for oil. The 1980s and the 1990s were characterised by an increase in the production of the non-OPEC oil on account of new discoveries.

As a consequence, OPEC's attempt to control was a not so successful. The fear of revenue loss among the OPEC countries prompted them to violate the quota often. Even Saudi Arabia began exceeding the quota limits from 1985, bringing down the price level to below $10 by 1987. The 1991 Gulf war only led to a temporary price hike. Only in the 1994-1997, rapid economic growth rates in the US and in the Asia-Pacific regions led to somewhat price recovery. The economic crisis in the Asia-Pacific region led to the actual decline in the oil consumption in Asia in 1998 leading to the plummeting of the price in 1998. The link between the economic recovery and the oil price rise is not so strong since the early 21st century.

Future Thrust

There is an increased promise for oil use and trade for the next half a century: especially because lot of oil is found and is yet to be explored in Russia, the Caspian region, the West Africa, the North America and the South America.

Russia already is an energy superpower. To achieve its full potential, Russia needs to strengthen corporate governance and the legal/regulatory framework for business, improve its foreign investment climate, allow competition in the transportation system, open the gas and oil companies: Gazprom and Transneft up to reform and competition, improve its technological capabilities, and move domestic energy prices to world levels. The Caspian Basin has tremendous potential. Upto 5.0 million b/d could be produced.

The key issues in the Caspian energy development include:

1. Completing the second pillar of the East-West Energy Corridor by developing the South Caucasus natural gas pipeline.
2. Improving the investment climate throughout the region and
3. Bringing Kazakhstani oil into the East-West corridor.

Africa is playing an increasingly important role as an energy supplier to US and to the world at large. In 2003, Nigeria and Angola were among the top 10 suppliers of oil to the United States. Oil production generates substantial revenue in countries such as Nigeria, Angola, Gabon, Equatorial Guinea, Republic of Congo, Chad and Cameroon. São Tomè and Mauritania also may become oil suppliers in the coming years. Higher foreign direct investment is needed to develop African energy resources. Most of the new fields are in deepwater offshore environments that require advanced capital-intensive facilities for development.

In spite of the increasing role of the non-Middle Eastern regions in the world oil market, the role of Saudi Arabia and the Middle East is indispensable. The Middle East not only holds 2/3rd of the world's oil reserves, but also has the lowest extraction cost of oil. The Middle East has got a lot to offer in the arena of international relations. While the region exports oil and has the capability to make huge investments in countries like India, it has a lot to gain in the form of technological inflows from the developed countries and from the countries like China and India.

Natural Gas Markets

Between 2002 and 2007, natural gas consumption in the non-OECD countries grew more than twice the rate of growth recorded in the OECD countries. Non-OECD countries would account for more than 90 per cent of the growth in world natural gas production between 2004 and 2030.

Worldwide consumption of natural gas would increase from 100 trillion cubic feet in 2004 to 163 trillion cubic feet in 2030. Natural gas is a key fuel in the electric power and industrial sectors. In the power sector, natural gas is an attractive choice for new generating plants due to higher fuel efficiency, clean combustion and very low carbon emission.

Industrial sector accounted for 44 per cent of world natural gas consumption in 2004 and is projected to account for 43 per cent in 2030. Natural gas use in the industry would grow at CAGR of 1.9 per cent while the growth of liquid fuels would be at a CAGR of 1.1 per cent.

Reserves and Resources

The proved world natural gas reserves on 1 January 2007 were estimated to be 6,183 trillion cubic feet—71 trillion cubic feet (about 1 per cent) higher than the estimate for 2006. Kazakhstan, Turkmenistan and China experienced a huge net increase in the natural gas reserves with the US also reporting notable increase.

Almost three quarters of the world's natural gas reserves are located in the Middle East and Eurasia. Russia, Iran and Qatar together accounted for about 58 per cent of the world's natural gas reserves. The reserves in the rest of the world are fairly evenly distributed across the regions. The reserves-to-production ratio worldwide is estimated at 65 years. The ratios for Central and South America, Russia, Africa and the Middle East are about 52, 80, 88 and over 100 years respectively.

Natural Gas Supply

As noted above, the non-OECD countries would account for about 90 per cent of the growth in supply between 2004 and 2030 and CAGR would be about 2.6 per cent. Russia and the Middle East would have a market share of about 20 per cent each with extensive pipeline networks catering to the dominant and industrialised markets.

Russia has an extensive pipeline network reaching into Europe and has proposed the construction of pipelines into China and South Korea. Russia is also exporting LNG from its Pacific coast to the Atlantic market. Middle East exports significant quantities of LNG to the Atlantic and the Pacific basins. In 2005, 15 per cent of the LNG exports from the region went to North America and Europe and 85 per cent to Asia. Qatar is becoming the major source of LNG imports into North America and Europe.

Africa and non-OECD Asia (excluding China and India) would become important sources of natural gas production. From 14 per cent in 2004, the combined global share of Africa and the non-OECD Asia would increase to 21 per cent by 2030. The export share was 26 per cent and 50 per cent respectively in 2004 for non-OECD Asia and Africa and would decline to 10 per cent on account of increased domestic consumption in non-OECD Asia and would increase in Africa by 2030.

In the case of North America, the US would increase the diversification of natural gas imports away from Canada and Canada's domestic production of conventional gas in the Western Sedimentary Basin is declining.

Canada's share of US' imports would fall from 85 per cent to 25 per cent between 2004 and 2030. The hitherto uneconomical gas pipelines *viz.*, the Mackenzie Delta pipeline and the Alaska pipeline are becoming economical and would account for a significant portion of the US' natural gas imports.

A large portion of North America's technically recoverable resource base comprise unconventional sources including tight sands, shale and coal bed methane. Tight sands formations are expected to be a significant source of US' incremental natural gas supply, increasing from 40 per cent of total domestic production in 2004 to 50 per cent in 2030 and accounting for 28 per cent of the increase in US' natural gas supply.

Demand for Natural Gas

OECD North America

Natural gas consumption in North America is projected to increase at an average annual rate of 1.0 per cent from 2004 to 2030. The average annual growth rate of natural gas demand in US is projected to be 0.6 per cent, significantly less than that in Canada and Mexico, largely due to the impact of higher natural gas prices and supply concerns in the US natural gas markets. US' share in the North American natural gas market is expected to fall from 80 per cent to 74 per cent between 2004 and 2030.

OECD Europe and Asia

At a CAGR of 1.4 per cent, natural gas will be the fastest growing fuel in OECD Europe and Asia. Most of the natural gas growth is due to the quest for less carbon intensive power generation. Especially in Japan, the CAGR for power generation would be 1.7 per cent.

South Korea tends to drive up the natural gas prices due to lack of underground storage facilities and exclusive reliance on imports: especially during the winter due to the need for residential heating.

Non-OECD Europe and Eurasia

The region's reliance on natural gas is highest with 55 per cent of Russia's energy requirements being met by the natural gas. On the whole, 44 per cent of the region's energy requirement is catered by the natural gas.

Non-OECD Asia

The region should experience the fastest growth rate accounting for 30 per cent of the global growth. The consumption would triple between 2004 and 2030 and CAGR led by China and India would be 4.6 per cent. The share of natural gas in China's and India's fuel mix was only 3 per cent and 8 per cent in 2004 and would at least double by 2030.

India is also driving up the natural gas price by paying $9 per million Btu although not so much as the South Korea which would end up paying $26 during the winter season. India's new gas fields would serve existing as well as the new demand for natural gas by 2015 and would cater to a CAGR of 6.2 per cent until 2015.

Natural Gas Pricing

Qatar has pegged its LNG sales to crude oil prices in Japan, to Henry Hub spot prices in the US, to National Balancing Point (NBP) spot prices in the UK and to fuel oil prices in continental Europe. In Japan, cif[1] LNG prices are based on a basket of imported crude oils, known as the Japanese crude cocktail (JCC), which are adjusted on a monthly basis.

In the past, this 'cocktail' was a convenient basis for gas pricing because the main competitor of gas was light crude oils, whose prices are reflected in the JCC. However, as crude oil no longer competes with gas in the electricity generation, Japanese buyers require more flexibility in the volume and pricing of natural gas.

The contract signed by Australian North West Shelf (NWS) with Chinese buyers has set a new benchmark for LNG pricing in Asia. The price formula is reported to be similar to that used for the NWS Japanese contracts *viz.*, an S-curve formula based on the average price of a cocktail of

1. cif - Cost insurance+freight (average prices).

imported crude, designed to protect the parties against sharp swings in oil prices. However, based on a reference barrel of $18, NWS partners have lowered the cif price to China to around $3/million Btu, or about 25 per cent below the current Japanese price. Moreover, the slope of the Five-curve is not as steep as under the Japanese formula. This means that for a price of $25/barrel, China would pay 25 per cent less than Japanese buyers. This price cut will greatly affect renegotiation of the Asian contracts which come up for renewal by 2010 (about 25 Mt including Japanese contracts with NWS partners).

In Europe, pooled electricity prices along with the heavy oil price serves as a reference point. Hence, periodic review and renegotiation of the contracts is possible unlike in Japan.

The US' Henry Hub's pricing reflects the distance of the consumer from the LNG terminal and allows for an 80-90 per cent shift of the spot price *vis-à-vis* the shift in the futures price of natural gas. Naturally, expansion of the pipelines would partially restore the Henry Hub spot prices in the event of a large influx of LNG imports leading to a price crash. The Henry Hub formula limits the scope for arbitrage and futures trading in natural gas.

In the Pacific Basin especially, such arbitrage is not possible as long as no re-gasification capacity is available on the US West Coast or Mexico. Even when such capacity becomes available, large transportation distances would limit the possibility of arbitrage compelling the markets to clear on the spot locally.

The evolution of the LNG markets is making the Middle East play a swing role or the role akin to the OPEC.

Main Factors Affecting the Price of Natural Gas and LNG

1. *General situation of the country:*
    - Whether it is gas-short or gas-rich country.
    - Sharing of risk.
    - Share of the gas compared with the competing fuel.
    - Penetration strategy and penetration speed.

2. *The quality of the product:*
   - Reliability of gas.
   - Flexibility of deliveries.
   - Quality.
3. *Market conditions:*
   - Which competing fuels?
   - Shares of the competing fuel.
   - Changes of the competing fuels over time or by price development.
   - Any premium may be subject to change or by price development.
   - Changes in the market structure.

The price will account for the extraction cost along with the transportation and distribution costs. Usually, 'netback value' *viz.*, the maximum price of gas at the consumption point excluding the intermediate costs *viz.*, re-gasification, transportation, liquefaction etc., is used.

The Principle of Gas Pricing

The issue is controversial enough.

1. Setting producers prices high enough to tempt the risk takers into exploration and production.
2. Creaming of some of the resources (rent) for the government.
3. Finding equitable transfer prices at the producers and the utilities level.
4. Devising complex tariff structures to reward large consumers for shifting to off-peak period.
5. Protecting small consumers from monopoly exploitation by the utility.

The whole chain of prices, from the producer to the smallest consumer must reflect and properly distribute the economic gains from the development of gas.

Objectives of Gas Pricing

Prices set by the central authority are generally based implicitly or explicitly, on three sets of objectives:

1. *Efficiency of resource allocation:* The economic efficiency of pricing should reflect the real cost of production and usage of energy.
2. *The satisfaction of specific financial targets:* The tariff structure should lead to financial solvency as well as to the efficient use of energy.
3. *Consideration for social equity:* This objective would contradict the previous method because the subsidies are a direct contrast to the financial viability of the project.

The Level of Gas Prices

Economic efficiency should be the primary criterion for setting the gas price. At the upstream gas price is to be determined by base prices, indexation, floor prices, capacity charge (load factor), currency adjustment, lag periods and price preview clauses. Gas pricing indexation yardsticks differ since they are established by bilateral agreements.

Pricing Formula

LNG is usually sold on the basis of pricing formula that usually links its pricing to the price of the crude oil.

The formula is usually structured as:

$$\text{Price paid on day one} = \frac{\text{Base price} * (\text{MPI-1}) + \text{A}}{\text{MP}}$$

Where, MPI-1 is the market price of specified cocktail of crude oils imported on a specific day.

MP is the market price of the specified cocktail of crude oils imported on a day of the contract.

A is an adjustment factor often specified with limits.

LNG Pricing in the Arabian Gulf Region

Two Distinct Formula are Used in Arabian Gulf Region

Qatar's pricing formula in contracts with Japan calculates LNG price by taking the Japanese custom cleared price of a basket of crude and multiplying with a factor of 0.1845 or $6.73 mmBtu/bbl of oil. A fixed factor is then added to the price which is between $ 0.6 to $ 0.9 mmBtu is meant to account for the relatively fixed transportation costs for the gas.

Oman's price formula calculates LNG prices on the basis of Japanese customs clearing price of the crude oil on a free on board (FoB) basis to be multiplied by the factor of 0.1515 or a conversion factor of 6.6 mmBtu/bbl of oil. It neither contains a fixed element nor specifies a minimum price.

LNG Pricing in Europe

The European gas industry witnessed stable and steady growth for nearly four decades and is undergoing a profound change. Restructuring of the gas market is leading to the emergence of new players. Third party access rules are being promulgated for LNG import facilities and natural gas transmission. In the era of increased competition, price would largely determine the success of the market players. In the absence of the international benchmark price unlike in the case of oil, the spot natural gas price is linked to the price of the competing fuels.

Long-term contracts predominate in Europe and the price is based on two factors.

- The base price, which splits the risk between the two parties, sellers and buyers.
- Indexing, this ensures gas competitiveness and helps to integrate the changes occurring in the energy market, without having to renegotiate the contract.
- European gas price has been based on the netback market value approach. The price paid by the importing company to the producer at the border or beach is negotiated on the basis of the weighted average value of the gas in competition with other fuels *viz.*, gas-oil,

low sulphur, high sulphur, heavy fuel oil etc., in the end-user market adjusting for net taxation and transportation and handling cost.

The formula is:

$Pt = Po [a (Gt/Go) + b (Ft/Fo)]$

Where:

Pt is the price of gas at time t,

Po is the price of gas at time o,

Gt is the price of gas oil at time t,

Go is the price of gas oil at time o,

Ft is the price of fuel oil at time t,

Fo is the price of fuel oil at time o.

Since LNG competes directly with pipeline gas in the end-user market, LNG prices are indexed to a similar mix of product prices. Due to the indexing, European LNG and pipeline gas prices are generally lower than crude oil prices.

In recent times, short-term market has resulted in the growth of spot market that determines the price on a day, or a week, or a month or even on a year ahead basis. The price is negotiated at the end-user level which minimises the hedging risk under the NBP trading scheme in the UK.

LNG Pricing in Asia-Pacific Region

In the Asian market, 97 per cent of the gas is LNG of which, Japan imports over 50 per cent of the world production. The Asia-Pacific LNG prices are indexed to crude oil prices as manifested in Japan and in Indonesia.

Specifically, the JCC prices which is at times restricted to limit the impact of extreme oil price movements. JCC is the average price of the crude imported into Japan every month.

The traditional formula is:

PLNG = A * P crude oil + B,

PLNG = price of LNG in cents/mmBtu,

A = slope of line showing crude linkage,

P crude oil = price of crude $/bbl,

B = a constant in cents/mmBtu.

LNG Pricing in USA

Based on the Henry Hub price which was discussed earlier, the price of the gas at various points in the continental US is derived from the differentials (+1-) from the Henry Hub. The differentials are based on demand patterns and distance of end-users. The Henry Hub pricing allows greater degree of transparency in the short as well in the long-term contracts.

LNG Economics

The typical investment needed for setting up a LNG import and export facility with a capacity of 5 million metric tonnes per annum (Mmtpa) is estimated at $2.5-3 billion which is equivalent to $1.5-2 billion needed for the gas treatment and liquefaction plant, $500 million for two LNG carriers and $500 million for an import terminal and re-gasification plant. Hence, the entire chain of natural gas should be accounted for by the project.

The cost of LNG would be higher than the cost of domestic gas because LNG involves significant processing costs like liquefaction, transportation and re-gasification in addition to the cost of production.

To price LNG competitively, the entire cost chain has to be analysed.

Table 5.4

Typical Capital Costs of 2.5 Mmtpa LNG Project

	Capital Cost (US$ million)	*Cost Per Unit Gas (US$/MmBtu)*
Upstream	400	0.60
Liquefaction	750	1.10
Shipping	500	0.6
Re-gasification	400	0.6

Source: TERI Volume on Energy Trading 2008.

Table 5.5

Capital Costs for Generic LNG Liquefaction Terminal

LNG Liquefaction	*2.5 Mmtpa*	*5 Mmtpa*
Liquefaction, utilities, storage etc.	460	921
EPC	103	154
Marine	28	28
Infrastructure	47	47
Owners	112	112
Total	750	1262
Cost date	for typical LNG ship	
LNG tanker		
Vessel size (cum LNG)	138.0	
Ship speed (knots)	17	
Capital expenditure ($ million)	250	
Operating costs ($ million)	6 per annum	
Insurance	1 per cent capital expenditure	

Note: EPC: Engineering, procurement and construction.

Table 5.6

Capital Costs for Generic LNG Receiving Terminal

(in $ million)

	2.5 Mmtpa	*5 Mmtpa*
Unloading marine	49	49
Storage, vaporisation, utilities etc.	148	268
Fuel oil storage, building, facilities	34	135
Direct field costs	231	352
Field distributable	31	44
Field costs	262	396
Project mngt, engineering + contingency	137	190
Total	892	1434

The operating and maintenance and insurance costs for a liquefaction plant are assumed to be 2.50 per cent of the capital cost. Significant economies of scale can be realised by optimally using the fixed infrastructure like the plant and by strategically increasing the use of quasi-fixed costs like the transportation and the storage infrastructure along with the variable factors.

The secret behind cost lies in the sizing and scaling. Transportation is the key factor either resulting in cost increase or cost reduction. For example, a large ship can transport LNG across great distance instead of two large boats. The overall capital cost for LNG projects are enormous and such projects require enormous skill and resources in matching each link of the chain together.

An LNG project chain is supported by the end-user contract, which is a gas sales agreement with the off taker who in turn enters into power purchase agreement with the customers. The long payback period necessitate the long-term contracts along the LNG capital chain for 20-25 years. Due to the high debt service requirements associated with the financing of the various project components, 'take or pay' contracts are entered into. Under these contracts, the off taker pays for the predetermined quantity whether sold or not. There are devices to differ the payment by the off taker *viz.*, the makeup delivery contracts.

According to a study conducted by Asian Development Bank, the delivered cost of LNG varies between $2.55/mmBtu and $3.65/mmBtu depending upon the terminal and sources used. LNG can be delivered most cheaply to the west coast of India from the Gulf, the net backed pricing calculation indicates that the imported gas could be delivered and re-vapourised at less than $3.25/mmBtu.

Supplies from Malaysia would be landed at a similar price at the East Coast. Gas from Australia would be relatively more expensive due to substantial additional shipping cost involved. The closest source of LNG to the West Coast and Southern India could be from Arun in Northern Sumatra, Indonesia. If a swap arrangement could be enacted, such that LNG from another Southeast Asian source was substituted for Arun LNG to Japan, revapourised LNG could be supplied to Gujarat at a higher rate than to other locations due to the higher costs involved in constructing the appropriate marine facilities.

Concluding Remarks

Oil has a role to play for another 40-60 years and natural gas 60-100 years respectively depending on the estimates. The pressing concerns faced by the world are the sustained increase in the production and supply of oil and natural gas. Technological innovations are and should take place at the wellhead to make the hitherto uneconomical oil resources more economically extractible.

The supply lines should be secure from geopolitical risk of attacks and vandalisation. Suppliers should have the economic incentive to supply oil and natural gas. Therefore, long-term price contracts should encompass all the possible costs and risks. At the same time, the pricing of oil and natural gas should not be prohibitory of economic activity.

In the case of India, a sustained 8-9 per cent growth rate in the economy should be maintained until 2032 and 6-7 per cent until 2050-2060 to overcome poverty and educational gap. Strategic trade ties should be established by India in the East Coast and in the West Coast to tap the natural gas resources of Indonesia, Malaysia and Qatar at economically competitive price. India is attempting to overcome the foreign exchange cost of oil and natural gas import by trying to open up the refineries for export processing thus, contributing the much needed foreign exchange.

To establish geopolitical security of oil and natural gas, the US/Middle Eastern ties and the Russian/European and the Indo-Russian ties are being strengthened.

A limited amount of taxation is being enforced to collect funds for renewable energy research that smoothens the transition to renewable energy on an economic basis.

6 Power Trading

Introduction

The present chapter focusses on the trading of the power. Hence, the term 'power' replaces the term 'electricity' for the most part in order to ease the reader's concentration. Bulk of power transmission and trading takes place through the alternate current (AC) mode. There is a need to account for and regulate the generation, transmission, distribution and retailing of the power sector.

The demand for power is derived from its requirement in operating the domestic appliances, office equipment, industrial machinery and providing sufficient energy for both domestic and commercial lighting, heating, cooking and industrial processes. Due to its all encompassing nature, power is viewed as a 'public utility' and the infrastructure expenditure has to be significantly borne by the Government.

The chapter has been divided into 13 Sections. Section II discusses the relevance of power sector reforms. Section III discusses the types of power markets *viz.*, the wholesale and retail, and discusses the modes of pricing and operation. Section IV discusses the mechanism of power trading *viz.*, the ways of ensuring economic efficiency and supply security. Section V discusses the key drivers of power trading *viz.*, the transmission mechanism, balancing mechanism, free trade etc. Section VI introduces the concept of power being a commodity which is a step towards introducing the instruments for trading in power. Section VII introduces various physical and financial derivatives used in power trading (over the counter as well as in a power exchange) such as the forwards, futures and options. Section VIII discusses various types of options such as the spark spread, plain call or put, swing etc. Section IX discusses the structured bilateral contracts such as the tolling contracts and the load-serving full requirement contracts. Section X discusses the financial derivatives on the power transmission capacity such as the Financial Transmission Rights and the Flowgate Rights. Section XI discusses the ways of pricing the power derivatives such as the fundamental approach based on the realistic market scenarios and technical approach based on the capacities of various compartments such as the generation, transmission and the distribution infrastructure. Section XII discusses the market models in vogue in various leading countries such as, the UK, the USA, the Nordic pool and India. Section XIII presents the concluding remarks.

The chapter also has an appendix illustrating the functioning of various types of power markets such the UK and the NYMEX.

Power Sector Reforms

Power industry is concentrated in many countries. In other words, the generation, transmission, distribution and retailing are owned by a single utility. The industry is generally regulated heavily and is often characterised by price controls. The industry is usually owned and operated by the Government. Reform aims at increasing the competition and breaking up the 'natural monopoly' in the ownership of the infrastructure. Reforms also tend to facilitate the people's ownership of certain portion of the infrastructure.

Reforms are bringing in many power generation companies which compete in selling the power to the distribution utilities and to the retail dealers who also compete to purchase the power. Alternately, the modern concept of the 'power exchange' enable the competing power generation companies to sell the power to a pool, the access to which is determined by the competition between the brokers, distributors, retail traders and the end-use customers.

The demand for power is linked with the willingness to pay. Reforms or the deregulation is expected to enhance private investment, increase efficiency, promote technical growth and improve customer satisfaction by virtue of increased competition among the producers, traders and the consumers of the power. Winning the market share and remaining in the business are the most important goals of the companies under the new reform or deregulated structure.

Free and fair power market is ensured by the open access regime. Under this regime, a common transmission network is used by the companies and the customers to transmit the power from the point of generation to the point of distribution and perhaps to the point of end-use. The transmission network is thus considered a 'natural monopoly' that helps avoid the problems of duplicity of the infrastructure and right-of-the-way. This in turn contributes to reduced installed capacity, increased system reliability and improved system performance.

Deregulation of the power sector is meant to improve the rationale and bring in transparency. This also helps in determining the exact quantum of reserve power to improve the system reliability. Reserves are either produced on demand or are imported on purpose.

It is not important for us to associate the deregulation with privatisation. What is important is weather the whole sale and retail markets are competitive and weather the generation and the retailing functions are firmly established on a competitive footing *vis-à-vis* the natural monopoly of the transmission and distribution.

Types of Power Markets

Wholesale Market

In a wholesale market, retail dealers and the financial intermediaries purchase power from the generation companies on a spot or forward basis. In the spot market, the currently prevailing price is paid. In the forward market, the power is agreed to be delivered at an agreed point of time at the predetermined price.

Power cannot be stored. The demand and supply of the market has to be cleared on the spot. The controlling role is played by the transmission agency which sends the information to the generator and the distributor concerning the expected demand and supply. If demand exceeds the supply in very short time *viz.*, the next hour or so, the frequency would be increased by about 50 or 60 hertz (Hz). If necessary, power generation itself could be increased. If vice-versa, either the frequency is reduced by about 50 or 60 Hz or the generation is curtailed. Apart from the demand, the generation of power is determined by the extent of the Transmission and distribution (T&D) losses or the extent of the congestion in any part of the network.

Competitive power market is characterised by the bid-based, security-constrained, economic dispatch with nodal prices. The theoretical price of power at each node on the network is a calculated 'shadow price'. It is assumed that an additional kilowatt-hour is demanded at the node in question, and the hypothetical incremental cost to the system due to the re-dispatch of the available units is calculated based on that shadow price that also includes the T&D losses and the costs of network congestion.

Since the price is based on the distance of the distribution or the end-user node from the generation point, the price in question is known as the 'locational marginal price' or the 'LMP'. Most operators with the exception of the UK and the Nordic pool comprising the Scandinavia and Finland do not use this pricing. Many reserve units supplying power to meet the anticipated and sudden demand are compensated in accordance with their cost and thus the actual cost of power is not matched by the price paid by the consumers. Political quest for a uniform affordable price for the consumer is the reason behind the below cost pricing.

In spite of the deregulation, only the short-term efficiencies in terms of the load management and dispatch systems are improved by the transmission utilities. Below cost pricing acts as a barrier to stimulate new entrants and new investments towards the expansion and modernisation of the power sector. However, market determines the shortage in the foreseeable future and power cuts instead of being imposed, become voluntary as in the case of New Zealand which experienced severe hydro-power shortage on account of severe droughts between 2001 and 2006.

Market functioning based on the LMP enables the prediction of the thermal losses due to the generation constraint. However, market fails to predict the 'security constraint' or the failure in the functioning of a power line or a section of it. However, the new computerised technology enables the monitoring and price calculation on a 'day ahead' basis and even prior to 'half an hour.' The spot market should take care of the unplanned outage of the power line. Thus, the 'bid based, security constrained economic dispatch and model price' model becomes applicable in the spot market.

The following are some of the important wholesale power markets:

- Australia—The National Electricity Market Management Company Ltd. (NEMMCO) the Australian market administrator.
- Canada—The Independent Electricity System Operator (IESO), Ontario and Alberta Electric System Operator (AESO).
- Chile.
- Scandinavia—The Nordic Power Exchange(Nord Pool).
- France—Powernext.
- Germany—European Energy Exchange (EEX).
- Great Britain—Elexon.
- India – The Power Trading Corporation of India Ltd.
- New Zealand—New Zealand Electricity Market.
- Philippines—Philippine Wholesale Power Spot Market.
- USA—Electric Reliability Council of Texas (ERCOT) market, PJM (Pennsylvania, New Jersey, Maryland) market, New York market, Midwest market and California Independent System Operator (CISO).
- Singapore—Energy Market Authority (EMA), Singapore and Energy Market Company (EMC).

Retail Power Market

In a retail power market, the end-use customer can choose the supplier. There is intense competition between the retail suppliers in the developed markets. Consumers usually are unwilling to face the 'real time' price or the price variation based on the changes in the wholesale price. Hence, the retail pricing would be based on the average annual price so that demand variation does not occur in random with the variation in the wholesale price. At the most, price variation strategies could help reduce the peak demand or help implement certain technical solutions.

It is possible to have several retail suppliers alongside one wholesale transmitter. However, the node on the transmission grid and the power quantities at that node has to be reconciled so as to facilitate the competition for retail customers beyond the node but within the distribution system. In the German market for example, large, vertically integrated utilities compete for customers on a more or less open grid. Although the market structures are different, power retailers have to perform certain common functions or enter into contract in order to compete effectively. Failure or incompetence to execute one or more of the following had caused catastrophic financial disasters:

- Meter reading,
- Meter rental,
- Billing,
- Credit control,
- Customer management via an efficient call centre,
- Distribution use-of-system contract,
- Reconciliation agreement,
- 'Pool' or 'spot market' purchase agreement and
- Hedge contracts—contracts for differences to manage 'spot price' risk.

Risk management and billing are the main areas of weaknesses so far. The California Power crisis of 2001 was due to high spot prices against which the incumbent retailers had no means of hedging against. In UK, a retailer 'Independent Energy' went bust when it could not ensure the payment of bills by its customers.

Power Trading Mechanism

Power traders should ensure the economic efficiency and supply security. Power traders have a two-fold mandate:

a. To develop a full-fledged, efficient and competitive market mechanism and
b. To facilitate the development of power generation projects which result in reliable, economic and quality power in the long term.

Power market should be vibrant. It should be deep and liquid. Power traders play an important role towards establishing such a market and subsequently add value to their customers through the provision of services meeting their evolving needs.

Power markets should ensure the optimal utilisation of the energy, to overcome inter-regional surplus-deficit and should serve as a catalyst to promote fresh investments, modernisation and environmentally-friendly power projects. If necessary, power should be exchanged between the neigbouring countries. Power cannot be stored and its supply cannot increase at the point of demand due to various constraints including the fuel and technology.

Due to the above constraints, there is a need to regulate demand by adopting the differential tariffs based on the time-of-day and the season. State Electricity Boards (SEBs) can shift the agricultural load to the nights. In medium- to long-term, capacity addition has to take place. While the SEBs managed to almost achieve their targetted capacity addition in the Ninth Plan, Central utilities and the private sector have failed miserably achieving only 37 per cent and 28 per cent of their respective targets.

India's normal power shortage is about 11 per cent and the peak time shortage is 15 per cent respectively. This is in spite of the fact that about 1,67,000 MW of the capacity is installed.

India's transmission and distribution losses are about a third of the total electricity produced. India's inter-regional power transmission capacity across the five designated regions (northern, southern, eastern, northeastern and western) is only sufficient to meet the national deficit. If the electricity cooperation with Bhutan and with Pakistan materialises, our electricity problems at the level of present and immediate future consumption would be solved. Risk sharing is being affected between the producers and consumers.

Power Trading Corporation of India Limited (PTC) was set up in 1999 to ensure power trading and to achieve economic efficiency and supply security. PTC should develop a full-fledged, efficient and competitive power trading market mechanism that results in reliable, economic and quality power in the long term. An integrated national power market is the core of the 'statement of purpose' of the PTC.

Rationale behind Power Trading

Optimum utilisation of existing resources would become affected with the deficit in one region made up by the surplus in the other. The reforms can be implemented more rigorously through regulations. Investment in the new generation capacity in the power deficit region can be deferred.

Average cost of power can be brought down by infusing competition and utilising the surplus power. Grid discipline would be enhanced by increasing the scheduled exchanges and by decreasing the un-scheduled exchanges.

Main Issues in Power Trading

a. Financial viability of the utilities: Operation of the divided utilities must be financially viable. The tariff structure, payment mechanism, payment before tax and payment after tax must be appropriate. This requires the balancing of the interests of the utilities and the consumers.

b. Power trading would ensure commitments by the utilities: Currently, utilities do not guarantee to provide adequate service. Disconnection is favoured over any other mechanism of payment realisation. Payment of power bills is not properly ensured. This results in the sickness of the utility, leading to its ultimate collapse.

c. Power trading should address the issue of adequacy of transmission capacity. High voltage direct current (HVDC) technology has not succeeded because the system has not reached the optimum capacity utilisation and hence is still costly to transmit power through HVDC. Inter-regional transmission capacity has to be augmented and wheeling charges have to be managed properly to ensure confidence among the utilities and the consumers.

d. Direct sale to the customers should be properly licensed and the issues of taxation and the cap on the margin should be properly fixed and monitored.

e. A well defined market mechanisms should exist: Power exchanges can serve the objective. India has two power exchanges which commenced their operations in 2008. The Power Exchange India Ltd. (PXIL) and the Indian Energy Exchange (IEX) are serving the needs of over 350 direct purchase customers. This customer base has to be augmented significantly. Exchanges should systematically redress the disputes.

f. Demand-supply mismatch should not hurt the interests of the customers. Surplus should be traded using the 'time of the day' or any other method. The prices should match the intrinsic value of power.

g. Grid discipline should be maintained for improving the quality of power (frequency excursions and voltages).

The above measures could bring down the average cost of power.

Key Drivers of Power Trading

Free Trade

All the opportunities of revenue generation should be tapped fully by the utilities. All the cost minimisation options should be fully explored and implemented to the best possible extent.

Transmission

Transmission is the most vital link between the power generator and the consumer. Age-old transmission links require augmentation. Inter-regional transmission capacity should be maintained optimally using any relevant technology *viz.*, the AC or the HVDC modes respectively. A national grid has been envisaged by 2012 with an inter-regional capacity of 30,000 MW.

Balancing Mechanisms

Balancing mechanisms like Availability Based Tariff (ABT) will enhance the intrinsic value of power and will reduce the average cost of power. Intermediaries like the power exchanges grow in numbers bringing in the necessary management and technical skills. State utilities would be able to negotiate long-term commercial contracts benefiting all the stake-holders. Surplus revenue can be utilised for augmenting the infrastructure and facilities to the consumers and the staff.

Open Access

Open access to the transmission and distribution network should be allowed with a definite timeframe. Cross-subsidy surcharge should be eliminated. Distribution, trading and transmission should be licensed.

Distributor should be allowed to trade, to franchise the distribution and generation and also to utilise the assets of other businesses. Cross-subsidy on open access and retail tariff must be phased out in a regulated manner to maintain system stability and to improve the health of the power sector. Central trading utilities (CTUs) and state trading utilities (STUs) must be barred from trading in power so as to ensure the open access.

De-Licensing of Thermal Power

Thermal power generation is de-licensed. This is the major source of power generation. Generation companies are free to choose their buyers. Captive power and co-generation are being encouraged with a greater emphasis on renewable energy resources. Regulator has a special role to play to oversee the operations of the market and apply appropriate controls whenever necessary.

Regulatory Risks

Unbearable tax burden and wheeling charges contribute to the regulatory risks. If not addressed, the projects can become unviable at the beginning or after a while. Payment security becomes difficult to enforce because bank guarantees may not be viable. Letter of credit is difficult to enforce. Cash transfer to the 'escrow' account would not materialise due to prior withdrawal by the purchasing utility or the consumer nor due to the bank's guarantees to meet other commitments through the internet banking mode requiring the automatic bill payment.

Power as a Commodity

Power cannot be differentiated and all the electrons are physically the same. Hence, power is defined as a commodity. Firstly, power is commodified at the wholesale level. The separation of the generation and the transmission functions enabled the germination of the wholesale markets. Novel contracts are being developed and some old ones are in vogue to facilitate the wholesale market. These contracts can either be sold on the bilateral market or on an organised market. They can either assume physical shape for delivery or financial shape *viz.*, the hedging. Three characteristics are common to physical and financial contracts *viz.*, a defined period, a certain amount of power and a price.

Physical Trading

Power can be bought and sold according to a variety of factors, such as how it is generated and when it is used. Utility companies can trade in the power to mitigate their price risks. Options such as the cost of generating their own power, purchasing power and selling power to smaller utilities and consumers to help mitigate their risks are valuable under relevant circumstances. A robust trading system is necessary to ensure a free and fair competitive market.

Trading system should hedge the risks associated with price volatility and other unexpected changes. Operating behaviour of a competitive power market is significantly affected by the trading arrangements, strategic bidding, market models and rules. Trading arrangement should be designed to prevent the abuse of market power. Transparency and non-discrimination are the core objectives of the trading arrangements which are frequently evaluated and altered to ensure the above.

Risk Management

Risk management is the very important component of the trading system. However, it is hard because power is not storable. In the competitive environment, the price is determined by the ever changing supply and demand pattern often that is mis-matched. Price volatility occurs almost on a daily and even on hourly basis. To hedge against price volatility, contracts are entered into.

Reservation Rights

Reservation rights are required in the event of the overloading or the congestion of the transmission line or the transformer that prevents the transmission of power from a generating unit. These rights are used to guarantee an efficient use of transmission system capacity and to allocate transmission capacity to users who value it most.

Physical Trading

Physical contracts can differ between long-term (1 to 20 years), forward (using current or forecast price with delivery in the future) and spot (very short term, mainly day ahead) ones. The necessity to ensure the supply/demand balance warrants long-term contracts. Inability to predict the actual supply/demand pattern warrant spot contracts. In most European countries, bilateral contracts are preferred pertaining to the long-term and forward contracts. Spot markets tend to be organised by power exchanges to overcome high transaction costs associated with bilateral contracts. Nordic countries, Germany, the UK and the Netherlands are the most important areas of power trading through the exchange in absolute volume.

Trading on a Power Exchange

Power spot prices are volatile due to the unique physical attributes *viz.*, non-storability, uncertain and inelastic demand and a steep supply function. Uncontrolled exposure to market price risks could lead to devastating consequences. During the summer of 1998, wholesale power prices in the Midwest of US surged to a stunning $7,000 per MWh as against the normal price range of $30-$60 per MWh. This caused the defaults of two power marketers in the East Coast.

In February 2004, persistent high prices in Texas during a three-day ice-storm led to the bankruptcy of a retail energy provider that was exposed to spot market prices. The California Power crisis of 2000-01 and its devastating economic consequences are largely attributed to the improper hedging of the major utilities through long-term supply contracts.

Power exchange provides a spot market (mainly day ahead) for power, which like any other market matches demand and supply for each hour, while providing a public price index. An exchange ensures competitive wholesale trading and is neutral to the market because its rules apply to both sides of the transaction. Power generators, distributors, traders and large consumers are the competing agents in the market.

The contract is bid based. Each sale bid specifies the quantity and its minimum price at which the generators are willing to supply power. Conversely, each buy bid specifies the desired quantity and the maximum price at which the distributors, traders or the consumers are willing to buy power, power exchanges help match the supply with demand and publishes a market-clearing price. Power exchanges are establishing their foothold in the pan-European wholesale spot market. The common goals of the power exchanges comprise:

a. facilitation of power trading,

b. fostering the development of competition,

c. insuring transparency and

d. Getting recognised as a European market place.

Finally, each power exchange aims to develop volume and the credibility of its price index.

Derivative Instruments in Power Trading

We discuss the various financial and physical instruments traded on the exchanges and over the counters (OTC). Most of the power futures and options on futures are traded on the New York Mercantile Exchange (NYMEX). However, the trading volume of power futures is less than power forwards traded in the OTC markets.

A large variety of power derivatives are traded among market participants in the OTC markets that include forward contracts, swaps, plain vanilla options and exotic non-standard) options like spark spread options, swing options and swap options. Other important trading vehicles for hedging the price risk of long-term revenue streams and service obligations are termed as structured transactions that include tolling agreements and load-serving full requirement contracts.

Derivatives

A derivative is a financial contract between two parties with opposite views on the market who are willing to exchange certain risks. Many derivative instruments are used in power trading. The most common ones applied to energy risk management strategies are future, forward and option contracts.

The plainest forms of power derivatives are forwards, futures and swaps. They are traded either on the exchanges or over the counters. These play the primary roles in enabling future price discovery and price certainty.

Power Forwards

Power forward contracts represent the obligation to buy or sell a fixed amount of power at a pre-specified contract price, known as the forward price, at certain time in the future *viz.*, maturity or expiration time. Power forwards are tailor-made supply contracts between a buyer and a seller, where the buyer is obligated to take and the seller is obligated to supply power. The price is based on the average price until the maturity date.

The peak time price of power on the maturity date is based on the average hourly price of the power daily until the date of maturity. Usually, the peak time is between 06:00 to 22:00 hours in the industry. Forward contracts are designed for the peak, off-peak and 'round the clock' markets. Generators such as independent power producers (IPP) are the natural sellers or, short side of Power forwards while LSEs such as utility companies often appear as the buyers or, long-side. Only contracts up to two years duration are traded.

Some of the forwards are purely financial contracts settled through financial payments based on certain market price index at maturity. The rest are physical contracts and are settled through physical delivery of power. Examples of financially settled power forwards include the contract for differences in the United Kingdom and Australian power markets.

Power forwards with short maturity periods like one hour or one day are often physical contracts, traded in the physical power markets such as the PJM power pool market and the energy balancing market operated by the California Independent System Operator (CAISO) in the US. Those contracts with a maturity period of weeks or months can be either physical or financial contracts and they are mostly traded through brokers or directly among market participants They are traded in the OTC markets.

Power forward contracts are the primary instruments used in power price risk management. LSEs *viz.*, the local distribution companies typically combine several months of forward/futures contracts to form a close' match with the long-term load shape of their customers. Other power marketers usually use forwards to hedge their positions in power options and in other complex power derivatives.

Power Futures

Power futures were first traded on the NYMEX in March 1996. They have the same payoff structure as power forwards. However, power futures, like other financial futures contracts, are highly standardised in contract specifications, trading locations, transaction requirements and settlement procedures.

The most notable difference between the specifications of power futures and forwards is the quantity of power to be delivered. The delivery quantity specified in power futures contracts is often significantly smaller than that in forward contracts. For example, a Mid-Columbia power futures traded on the NYMEX specifies a delivery quantity of 432 MWh of firm power, delivered to the Mid-Columbia hub at a rate of 1 megawatt per hour, 16 on-peak hours per day during the delivery month. A corresponding forward contract has a delivery rate of 25 MW per hour for the same delivery periods in a month.

Power futures are exclusively traded on the organised exchanges while power forwards are usually traded over the counter in the form of bilateral transactions. This fact makes the futures prices more reflective of higher market consensus and transparency *vis-à-vis* the forward prices. The majority of power futures contracts are settled by financial payments and hence lowers the transaction costs.

In addition, credit risks and monitoring costs in trading futures are much lower than those in trading forwards since the exchanges implement strict margin requirements to ensure financial performance of all the trading parties. The OTC transactions are vulnerable to financial non-performance due to counterparty defaults. Gains and losses of power futures are paid out daily. Power forwards cumulate and pay in a lump sum at maturity time.

Power Swap

Power swaps are financial contracts that enable their holders to pay a fixed price for the underlying power regardless of the floating power price or vice-versa over the contracted time period. They are typically established for a fixed quantity of power referenced to a variable spot price either at a generator's or at a consumer's location.

Power swaps are widely used in ensuring short to medium-term price certainty up to two years. They can be viewed as a strip of power forwards with multiple settlement dates and identical forward price for each settlement. Power locational basis swaps are becoming common and help to lock in a fixed price at a geographic location different from the delivery point of a futures contract.

A holder of a power locational basis swap agrees either to pay or to receive the difference between a specified futures contract price and another locational spot price of interest for a fixed constant cash flow at the time of the transaction. These swaps are effective financial instruments for hedging the basis risk on the price difference between power prices at two different physical locations.

Power Options

The emergence of the wholesale power markets and the dissemination of option pricing and risk management techniques in the post restructuring era *viz.*, from the early 1990s have created power options. The attributes like price, volume, delivery location and timing, quality and fuel type determine the nature of the options.

A counterpart of each financial option can be created in the domain of power options by replacing the underlying financial option with power option. Usually, the options are up to two years and the options with the time period greater than three years are long-term in nature and are known as the 'structured contracts'.

Plain Call and Put Options

Power call and put options offer their purchasers the right, but not the obligation, to buy or sell a fixed amount of power at a pre-specified strike price by the option expiration time. They have similar payoff structures as those of regular call and put options on financial securities and other commodities. The majority of the transactions occur in the OTC market. Power call and put options are the most effective tools available to merchant power plants and power marketers for hedging price risk because power generation capacities can be essentially viewed as call options on power, particularly when generation costs are fixed.

Spark Spread Options

Spar spread options are non-standard power options. Spark spreads are cross-commodity options paying out the difference between the price of power sold by generators and the price of the fuels used to generate it. The amount of fuel that a generation asset requires to produce one unit of power depends on the asset's fuel efficiency or heat rate (Btu/kWh). The holder of a European spark spread call option written on fuel G at a fixed heat rate K_H has the right to pay at the option's maturity K_H times the fuel price at maturity time T and receive the price of one unit of power.

Abstracting away the operational characteristics of a fossil-fuelled power generator *viz.*, the startup cost and ramping constraints, the per kW benefit of owning the right to use the generator is equivalent to having one kW spark spread call option with a strike heat rate matching the generator's operating heat rate. Based on this observation, it is clear that spark spread call options play important role in hedging the price risk of the output power of fossil-fuelled power plants and further serve as key instruments in valuing those generation assets.

Callable and Putable Forwards

Callable forward and putable forward are introduced to mimic the interruptible supply contracts and the dispatchable independent power producer contracts. In a callable forward contract, the purchaser of the contract longs one forward contract and shorts one call option with a purchaser-selected strike price. The seller of the forward contract holds opposite positions and can exercise the call option if the power price exceeds the strike price, effectively cancelling the forward contract at the time of delivery.

The purchaser gets an 'interpretability' discount on the forward price, equivalent to the option premium at the time of contracting continuously compounded to the delivery time. In a putable forward, the purchaser longs one forward contract and one put option with a seller-selected strike price. The seller holds the corresponding short positions. The purchaser exercises the put option if the power price drops below the strike price at the maturity time, effectively cancelling the forward contract. At the time of contracting, the purchaser needs to pay a 'capacity availability' premium over the forward energy price, which equals the put option price at that time, continuously compounded to the maturity time.

One variation of the callable forwards is proposed by adding an earlier notification date for exercising the call option in a callable forward before the contract matures. This emulates an interruptible service contract with early notification.

Swing Options

Power swing options are adopted from their counterparts in the natural gas industry. Also known as flexible nomination options, swing options have the following defining features.

Firstly, these options may be exercised daily or up to a limited number of days during the period in which the exercise is allowed.

Secondly, when exercising a swing option, the daily quantity may vary between a minimum and maximum daily volume. However, the total quantity taken during a time period should be within certain minimum and maximum range.

Thirdly, the strike price of a swing option may be either fixed throughout its life or set at the beginning of each time period based on some pre-specified formula.

Fourthly, if the minimum-take quantity of any contract period is missed by the buyer, a lump sum penalty or payment making up the seller's revenue shortfall should be paid.

Structured Bilateral Transactions

They enable the market participants to share and control a variety of risks including price and quantity risks over a potentially long time horizon.

Tolling Contracts

A tolling agreement is similar to a common power supply contract signed between a buyer *viz.*, a power marketer and an owner of a power plant *viz.*, an IPP but with notable differences. For an upfront premium paid to the plant owner, it gives the buyer the right to either operate and control the scheduling the power plant with the ISO or simply take the output power during pre-specified time periods subject to certain constraints.

In addition to inherent operational constraints of the underlying power plant, there are often other contractual limitations in the contract on how the buyer may operate the power plant or take the output power. For instance, a tolling contract almost always has a clause on the maximum allowable number of power plant restarts. These constraints make the pricing of tolling contracts a very challenging task. The analogy between holding a tolling contract and owning the underlying merchant power plant, however, leads to a numerical approach for valuing and hedging tolling contracts. Alternately, one may use a statistical approach for benchmarking the price reasonableness of tolling contracts based on historical power price and fuel costs.

Load-Serving Full-Requirement Contracts

Most large power consumers prefer a power supply contract with flexible consumption terms. Specifically, they desire to pay a fixed rate per unit of power for the actual consumption quantity, regardless of the quantity being high or low. Such a contract is termed as a load-serving full-requirement contract.

Suppose a power supplier (or, LSE) signs a full-requirement contract with a customer and then utilises futures contracts to lock in a fixed quantity of power supply at a fixed cost for hedging the expected energy consumption of the customer. The LSE is then at the risk of either under- or over-hedging, as the consumption quantity of the customer will almost surely deviate from the quantity hedged by the futures contracts.

When the power spot price is high or low, the total demand for power is likely to be high or low as well. A case in point is the periods of unusual cooling/heating needs. Hence, if the market price of power is higher than the fixed contract rate for serving power, customer's energy consumption level would be significantly higher than the hedged quantity.

As a result, the LSE is under-hedged relative to its load obligation and must purchase power in the open market to serve its customer at a loss because the wholesale spot price most likely exceeds the contracted price paid by consumers. Conversely, when the power spot price is low, the LSE faces the risk of being over-hedged and has to sell the surplus in the spot market or settle it financially at a price below its long-term contract price.

To hedge the volumetric risk, the LSE would need to buy a power option on the consumption quantity of its customers. Unfortunately, such an option is usually unavailable in the marketplace. Although perfect hedging may not be possible, weather derivatives that exploit the correlation between load and temperature can be used.

Financial Derivatives on Power Transmission Capacity

There are two major proposals for using financial instruments as transmission rights in the US:

a. the point-to-point financial transmission rights (FTRs) and

b. the flowgate rights (FGRs).

They are outlined in the Standard Market Design (SMD) put forth by the Federal Energy Regulatory Commission (FERC). FTRs and FGRs are power derivatives, with their values derived from the network transmission capacity.

FTR and FTR Options

In a power market such as the PJM that employs locational market price (LMP), a point-to-point FTR is specified over any two locations in the power transmission grid. An FTR entitles its holder to receive compensation (or pay) for transmission congestion charges that arise when the grid is congested.

The congestion charge/payment or the payoff associated with one unit of FTR is equal to the difference between the two locational prices of one unit of power resulting from the re-dispatch of generators out of merit order to relieve transmission congestion. The primary markets for the FTR trading are auctions held by the independent system operators (ISOs) of power markets. An FTR option offers the right to the FTR settlement without the obligation to pay when that settlement is negative. Hence, the settlement of an FTR option equals the positive part of the corresponding two-sided point-to-point FTR.

FGRs

Flowgates are defined over all transmission elements such as lines, transformers, or linear combinations of them. Each transmission element has two elemental flowgates, one in each direction. An elemental flowgate has a rated capacity in megawatts in its pre-specified direction corresponding to the capacity of an underlying transmission element. Thus, flowgate rights are link-based transmission rights for hedging transmission risks.

The values of flowgate rights can be established through auctions conducted by the ISOs. The spot price upon which the settlement of flow gate rights is based is given by the real time shadow price on the corresponding constrained element, determined by the security constrained economic dispatch algorithm employed by an ISO. Since these shadow prices are non-negative, FGRs are inherently defined as options.

Pricing Power Derivatives

Due to the unique physical and operational characteristics of power production and transmission processes, power price exhibits different behaviours than other financial prices. There are mainly two competing approaches to the problem of modelling power price processes:

a) Fundamental approach that relies on simulation of system and market operation to arrive at market prices and
b) Technical approach that attempts to model directly the stochastic behaviour of market prices from historical data and statistical analysis.

The first approach provides more realistic system and transmission network modelling under specific scenarios. it is computationally prohibitive due to the large number of scenarios required. Such analysis may be necessary for pricing financial transmission rights (particularly the flowgate rights) but not the other power derivatives.

Power Trading Market Models

A competitive power market should be supported by proper trading tools that take into account of special nature of power which is different from other commodities. A successful implementation of trading system in energy and its derivative markets could fulfill restructuring objectives, which include competition and customer choice and serve vital needs of power market participants.

Trading is an activity in which transactions take place directly between two participants or indirectly through an exchange.

Power trading through an exchange first commenced in 1996 in New York Mercantile Exchange (NYMEX). Power trading has two main components *viz.*, physical trading and financial trading. In physical trading, supply is balanced against demand and price is either determined in advance of trading or after trading. In financial trading, financial contracts take place between traders as agreements that give certainty to traders. Physical trading is generally done through an energy spot market or power pool. Financial trading is through a financial market or exchange such as NYMEX or Chicago Board of Trade (CBOT).

Power market trading is risky because the power is much different from other commodities due to its special nature. Power is non-storable, subject to generation-demand imbalance, characterised by limited demand elasticity and transmission constraints and price being related with the price of other volatile commodities. Power trading in the USA, UK and in some Latin-American countries underwent a profound change.

UK Power Market

UK is one of the leaders in developing spot power market. The initial competitive power market structure involved a power pool for England and Wales. The New Electricity Trading Arrangements (NETA) evolved as a result of the reforms.

The transactions taking place within the NETA market are power price-quantity transactions on a half-hourly basis. The system operator (SO) and power exchanges (PX) are central to the functioning of the E&W power market under NETA. PXs provide contracts that are 'cleared' giving a guarantee against default on a contract. Trading through brokers, OTC trading and other financial instruments (options and futures), also form the part of market trading. The SO for E&W, the National Grid Company (NGC), has two functions:

a. The management of high voltage transmission system as transmission system operator (TSO) and

b. The supply of operational services.

Generation forecasting, congestion management and provision of ancillary services are managed by NGC as SO. In addition, the balancing mechanism allows the SO to maintain local and national balances of generation and consumption in real time. Thus, the price and volume agreements within the balancing mechanism are made directly with the SO.

Nordic Power Market

Nord Pool, the Nordic Power Exchange, is the world's first international commodity exchange for power. The market trades with neighbouring countries and is dominated by hydropower. A transmission system links Denmark, Norway, Sweden and Finland facilitating the physical power exchanges. The national transmission system operators (TSOs) are responsible for reliability and balance settlements. Nordel facilitates cooperation between these TSOs and deals with planning, operation and transmission pricing.

Nord Pool organises trade in standardised physical and financial power contracts including the clearing services to its participants. The spot price for the Nordic power market is set by Nord Pool every hour. Elspot and Elbas are Nord Pool auction-based spot markets for trading the physical delivery power contracts.

On Elspot, hourly power contracts are traded daily for physical delivery in the next day's 24-hour period. On Elbas, continuous adjustment trading in hourly contracts can be performed until one hour before the delivery hour. New contracts are opened after the day-ahead Elspot prices are set.

Before 2 pm, the remaining hours of the current day are tradable and then day-ahead contracts are open for trading. The objective of Nord Pool financial market is to provide an efficient market, with excellent liquidity and a high level of security to the financial power contracts that can be used profitably by a variety of customer groups.

California Power Market

The competitive power market of California state of the USA commenced its operation in 1998 with the California Independent System Operator (CAISO) and now a bankrupt power exchange (PX) as the main operational market facilitators. The market took off smoothly, and the prices were seemingly just and reasonable until May 2000, when the first signs of market crisis emerged. California's power crisis was the result of the collusion of a shortage of resources, poorly designed market and inaction by regulators 'regulatory failure'.

The CAISO was originally designed to operate in conjunction with the PX, a day-ahead energy market that ceased to operate in January 2001. Without PX day-ahead market, all short-term balancing of supply and demand was thrust upon the most volatile real-time market. This is the result of flaws in original design and inconsistencies between the ISO's forward and real-time markets.

CAISO addressed these flaws through a comprehensive redesign effort known as Market Design 2002 (MDO2). It allows the ISO to match buyers and sellers through a transparent day-ahead market which reduces the reliance on the more volatile hour-ahead and real-time markets. Since the PX ceased operating, there has been no transparent market for spot power transactions to balance supply and demand ahead of real time. MDO2 also allows the ISO to manage congestion on the grid the day before, rather than in real time, thus enhancing real time reliability.

Indian Power Market

The Indian Power market is governed by the 'New Electricity Act' of 2003 and the 'Tariff Rules of 2006'. Free and competitive market and the open access regime are ensured in the acts and the rules.

Institutions in the Indian Power Sector

Central Electricity Authority (CEA)

The CEA is the Central government agency created under the Electricity (Supply) Act of 1948. It is responsible for regulating investments in the power sector.

State Power Boards (SEBs)

SEBs is the state power utilities. Every state in India has a State Power Board. Until the mid-1990s, the SEBs were the integrated power utilities. Orissa, Haryana and Andhra Pradesh pioneered the process of unbundling and establishment of separate corporations for the purposes of generation, transmission and the distribution of power.

Central Electricity Regulatory Commission (CERC)

The CERC was established by the Electricity Regulatory Commission Act, 1998. It consists of four members appointed by the Central government. The CERC shall be guided by such directions in matters of policy involving public interest as the Central government may give to it in writing. The CERC's aim is to promote competition, efficiency and economy in the power industry, encourage investment and safeguard the consumer interest.

The main tasks of the CERC are:

1. Regulation of tariffs of Central government stations;
2. Regulation of tariffs on electricity generated and sold across states in composite packages;
3. Regulation of interstate transmission tariff and
4. Regulation of interstate transmission;

In addition, the Act also gives the CERC an advisory role concerning electricity and the environment and to prepare and publish guidelines and arbitrate on certain issues.

State Power Regulatory Commissions (SERCs)

SERCs were also established under the same Act. The SERCs consist of at most three members appointed by the state government. The tasks of the SERCs are:

1. Determination of the tariff for power (wholesale, bulk, grid or retail);
2. Determination of the tariff payable for the use of transmission;
3. Regulation of power purchase and procurement process of the transmission utilities and distribution utilities and
4. Promotion of competition, efficiency and economy in the activities of the power industry.

Power Corporations (PC)

In the late 1970s, government established corporations like the National Thermal Power Corporation (NTPC), the Bharat Heavy Electricals limited (BHEL) and the Power Grid Corporation of India Ltd. (PGCIL). NTPC generates power from large pithead of the demand stations and sells it to various state utilities. NTPC satisfies about 22 per cent of the demand for power in India. BHEL and PGCIL are responsible for manufacturing of the equipment like turbines, boilers or transformers and for the erection and maintenance of interstate transmission lines.

Independent Power Producers (IPPs)

The efforts towards the privatisation of the power sector commenced in the mid-1990s with Orissa, Haryana, Uttar Pradesh and Andhra Pradesh being the pioneers.

Lanco Power, and its subsidiaries together is the largest private sector power producer with the total power generation capacity of about 3,300 MW closely followed by Tata Power with 2,971 MW.

Private power companies span all the areas of power generation *viz.*, the thermal, the hydal, small hydal, the renewable power etc. They have stakes in the transmission, distribution and retail trading networks. Managerial skills are the private sector strengths.

PTC India Ltd. is the leading provider of power trading services in India. It is trading in power since June 2002 through purchase from surplus utilities and sales to deficit State Electricity Boards (SEBs). PTC acts as a single-window service provider that manages both financial as well as operational risks for the buyers and sellers.

For its services, PTC charges a predetermined amount of transaction charge which is worked out on a per unit (KWh) basis or as a percentage of the cost of power. The price and margin information is known to both the counterparties in a transparent manner. PTC catalyses the development of power projects by entering into multi-year contracts for future trading of power. These power purchase agreements (PPAs) with PTC are being recognised by lenders as security for financial closure of power projects.

PTC is the nodal agency for cross-border power trading with neighbouring countries: Bhutan, Nepal and Bangladesh which are rich in hydropower resources. Utilities in Bhutan account for nearly 24 per cent of cross-border trading volumes. PTC operates an online power exchange since 2008 known as the 'Power Exchange India Ltd'. This apart, CERC has given a licence to another power exchange called the 'Indian Energy Exchange'. The performance of the power exchanges in terms of the volume of trade is miniscule. Out of over 800 billion KWH of power consumed in India, the power exchanges are trading far less than 5 per cent of the electricity.

Concluding Remarks

Power trading has come to stay. Various instruments of risk minimisation are being used to smoothen the flow of power and to hedge against the financial losses otherwise to be suffered by the utilities.

Power trading mechanisms such as the financial and physical instruments used in 'over the counter' and exchange-based trading are improving over time and the demand and supply are being tracked on a daily and even on an hourly basis. The electronic, communication and computer-based tracking and monitoring systems are enabling the timely detection of the problem, its repair and maintenance, ensuring proper and timely payment of the bills and almost instant communication between the power generation companies and its customers—including the large-end use customers.

In the case of India, power trading has taken off a decade ago. Over 80 per cent of trading takes place through the long-term contracts. The volume of exchange-based power trading is miniscule in India. However, there is a potential to enhance the exchange-based trading to manage the peak time power and to conduct 'load shedding'.

Exchange-based contracts are so far short-term in nature and are at the most 'week-ahead'. Only about 350 large customers are involved in the exchange-based trading.

Appendix 6.1

Illustration of Future Contracts

1. Power Contract on NYMEX

Trading unit: 40 megawatt hours (Mwh) per peak day (between 19 and 23 days) depending on the month. Depending on the number of peak days in the month, the number of megawatt hours will vary (between 760 Mwh and 920 Mwh).

Price quotation: US dollars and cents per Mwh.

Minimum price fluctuation: $0.05 per Mwh.

Maximum price fluctuation: no price limits.

Trading hours (all times are New York time): the contract is available for trading on the NYMEX ClearPort® trading platform from 6:00 pm Sundays through 5:15 pm Fridays, with a 45-minute break each day between 5:15 pm and 6:00 pm. Off-Exchange transactions can be submitted solely for clearing to the NYMEX ClearPort® clearing website as an exchange of futures for swaps (EFS) or exchange of futures for physicals (EFP) transaction until 5:15 pm, Monday through Friday, and the day preceding a holiday.

Trading months: the current year plus the next five calendar years. A new calendar year will be added following the termination of trading in the December contract of the current year. Peak days: Monday through Friday, excluding North American Reliability Council (NERC) holidays.

Termination of trading: one business day before the last peak day of the month.

Settlement: financial, based on the arithmetic average of the PJM western hub real-time locational marginal price (LMP) for the peak hours of each day provided by PJM Interconnection LLC. At the close of trading each day, starting with the last business day of the month prior to the contract month, 40 Mwh (a day's equivalent) will be subtracted from the quantity of the contract for the balance of the contract month.

For peak days that are not also business days, cash settlement will take place on the business day immediately preceding these days, separately, in chronological succession, if there is more than one such day in a row.

In the event a business day is not also a peak day, no cash settlement will be performed on the business day before. Cash settlement will be resumed on the business day that immediately precedes the next coincident peak day and business day.

Trading symbol: JM.

2. Ice Power Contract

Together with ICE's natural gas market, the availability of a power market on one platform provides an important advantage in speed and price information. More power is traded on ICE than any other electronic exchange in the world. ICE offers a broad range of futures and OTC products for base load, peak and off-peak periods, including products through 18 unique power hubs in North America and Europe. Both physically delivered and cash-settled power products are traded at a fixed price or differential to daily power indices. Power contract forms include swaps, futures, location spreads, time spreads and options.

ICE futures offers the UK power futures, and ICE's OTC markets offer trading in cleared or bilateral North American power hubs.

ICE OTC power: cleared and bilateral.

Eighteen (18) hubs—or delivery points—can be traded throughout the major FERC power regions in North America. Available both cleared and bilaterally, power contract forms include swaps, location spreads, time spreads and options. Hundreds of standard time strips are available, from same-day hourly power to full calendar years. Custom user-defined time strips are also available for some products.

Power products settle either financially in cash or with physical delivery at designated points and times. Both physical and financial power products can be traded at a fixed price or, for some products, at a differential to ICE's daily published power price indices.

3. UK Power Futures

The UK Power Futures contract is a deliverable contract offering all the benefits of exchange transparency and clearing in addition to the following key features:

- Base load and peak load contracts will be listed based on the EFA (Power Forwards Agreement) calendar and settlement cycle favoured by the industry with 12 months, 6 quarters and 4 seasons;
- No requirement to be a balancing and settlement code party;

- Margin offset between ICE UK power futures and ICE natural gas futures;
- Each contract will be physically deliverable and will be cleared by one central counterparty, LCH.Clearnet;
- Minimum trading size will be 10 lots;
- Months, quarters and seasons will be listed in parallel no cascading;
- All positions will be held as months for maximum flexibility for participants;
- There will be 50 per cent margin offsets between peak and base load contracts; inter-month spreads will be available and there will be price implication down the curve;
- Outright margin rates are envisaged to be £1.70 per MWh for base load contracts and £3.10 per MWh for peak load contracts.
- Intermonth spread rates are envisaged to be £2.25MWh for base load contracts and £3.75 per MWh for peak load contracts;
- The contracts will initially be available for trading through the website of ICE and will then be rolled to solutions.

Cessation of Trading: Month, quarter and season contracts cease trading at the close of business (17:00 hours London Time) on the second business day prior to the day on which power is to be delivered.

7 Nuclear Energy in India

Introduction

Nuclear power is small but growing fraction of India's energy requirements are likely to increase by ten-fold from 140,000 to 14,00,000 MW by the year 2050. The Government of India is ambitious to increase the proportion of nuclear power from 3 per cent to 30 per cent between now and 2050. Even by the year 2020, the proportion is likely to be increased to 10 per cent. India's nuclear capability has been growing during the past 60 years amidst great constraints and challenges. Foreign collaboration is hard to materialise. Even the cooperation from Russia was prevented during the post-USSR communist era due to pressure from the USA.

The recent cooperation between India and the USA under the ambit of the Indo-US nuclear deal can be attributed to the recognition of India as the growing economy, as a vast market for energy resources and a home for highly skilled and talented scientists and technologists. There is lot of policy, economic and strategic relevance behind this shift from non-cooperation to active engagement of US with India.

Engagement of the US with India opens the door for active cooperation of India with the 45-member Nuclear Suppliers Group (NSG) which will enable India to buy uranium from the international market. This enables the enhancement of energy security for India. However, there are challenges to the notion of energy security due to the finite quantity of raw uranium and the high expenditure involved in the reprocessing of spent uranium. These challenges can be overcome through the fast breeder reactor and the proposed thorium-based reactor.

The Indo-US Nuclear deal restricts India's ability to develop fast breeder reactors by imposing ban on reprocessing the uranium bought from

USA and uranium built with US' technology and collaboration. In this context, the discussion about India's policy and strategic independence with respect to nuclear power need to be discussed, ascertained and enacted upon.

In this backdrop, Section II of this chapter discusses India's nuclear policy *viz.*, comprehensive nuclear disarmament, credible deterrent, 'no first use of nuclear weapons' and self-reliance and the development of India's 'peaceful' nuclear programme in three phases *viz.*, the phase immediately after India's independence in 1947 until the first nuclear test in 1974, the phase between 1974 and the Indo-US nuclear deal of 2008 and the phase from the time the Indo-US nuclear deal itself came into effect. The section peruses the conditions and benefits of the nuclear deal.

Section III discusses the energy security, economic and strategic benefits arising out of India's nuclear policy with special reference to the Indo-US nuclear deal.

Section IV discusses challenges to India's nuclear programme in the context of foreign policy, geopolitical and strategic situations, the constraints these situations impose on India to fully comply with the Indo-US deal, US' strategic influence with the International Atomic Energy Agency (IAEA) and the possibility of overcoming the challenges through trade and investment policies encompassing the principles of joint responsibility and joint risk management.

Section V presents concluding remarks.

India's Nuclear Policy

When India emerged as a free country to take its rightful place in the committee of nations in 1947, the nuclear age had already dawned. India took the crucial decision for self-reliance and freedom of thought and action. Towards this goal, promotion of science and development of 'scientific spirit' was deemed essential by India's then Prime Minister Pandit Jawaharlal Nehru. India's nuclear policy was characterised by the quest for security and economic development.

Security

'Comprehensive nuclear disarmament,' 'credible deterrent', 'no first use' of nuclear weapons and 'self-reliance' are the corner stones of India's nuclear security.

Comprehensive Disarmament

Development of nuclear technology transformed the nature of global security. India reasoned that nuclear weapons were not weapons of war; these were weapons of mass destruction. The principle plank of India's nuclear policy states that a nuclear weapon-free-world would, therefore, enhance not only India's security but also the security of all nations. At the outset, India advocates nuclear weapons-free-world and non-discriminatory nuclear disarmament. Non-discriminatory choice of disarmament requires the exercise of freedom of thought and action.

In 1950s, nuclear weapons testing took place above ground and the characteristic mushroom cloud became the visible symbol of the nuclear age. India then took the lead in calling for an end to all nuclear weapon testing as the first step for ending the nuclear arms race. Addressing the Lok Sabha on 6 April 1954, shortly after the conduct of a major hydrogen bomb test, Pt. Jawaharlal Nehru stated that, 'nuclear, chemical and biological energy and power should not be used to forge weapons of mass destruction'. Pt. Nehru called for negotiations for prohibition and elimination of nuclear weapons and in the interim, a 'standstill agreement' to halt nuclear testing.

In 1963, an agreement was reached to prevent atmospheric testing of nuclear weapons. Since the five nuclear weapon states developed the technologies for conducting underground nuclear tests, the nuclear arms race continued unabated. In 1965, India along with a group of non-aligned countries put forward the idea of Nuclear Non-Proliferation Agreement paving the way for comprehensive nuclear disarmament if non-proliferation is adhered to.

In the final shape of the agreement, nuclear disarmament of the five existing nuclear powers was not included. Hence, India sought for 'nuclear umbrella' from the US or the USSR to remain safe from nuclear attacks. But, the nuclear umbrella was not extended to India. Hence, India did not sign the NPT keeping in view India's own self-enlightenment and national interest as stated by the then Prime Minister Smt Indira Gandhi.

Self-Reliance

India felt the need for nuclear weapons in the backdrop of two of the neighbours *viz.*, Pakistan and China acquiring the nuclear weapons. The first nuclear test in 1974 invited nuclear sanctions and the stoppage of international aid to India. The 1996 Comprehensive Nuclear Test-Ban Treaty (CTBT) did not have the option of nuclear disarmament. The NPT was extended indefinitely maintaining the gap between the nuclear haves and have-nots.

Credible Deterrent

The clandestine activities towards acquiring nuclear missiles, materials and technology increased in the neighbouring countries. India was subject to externally sponsored terrorism. Hence in the interest of enhancing the security through the enhancement of nuclear weapons capability, India did not sign the CTBT. India required the 'credible minimum deterrent' to thwart external aggression using nuclear weapons.

No First Use of Nuclear Weapons

Through various agreements, India expressed against 'no first use' of nuclear weapons. It expressed willingness to enter into separate pacts with countries towards this effect. India believes that 'no first use' agreement prevents the arms race. Historically, India's stand is justified by the past initiatives. In 1978, India proposed negotiations for an international convention towards the prohibition of the use or threat of use of nuclear weapons.

This was followed by another initiative in 1982 calling for a 'nuclear freeze'—a prohibition of production of fissile materials for weapons, on production of nuclear weapons, and related delivery systems. In 1988, India put forward an Action Plan for phased elimination of all nuclear weapons within a specified timeframe. India calls for Nuclear Weapons Convention along the lines of the Chemical and Biological Weapons Conventions along with some non-aligned countries.

India has even expressed willingness to phase out the existing stock piles of the chemical weapons. India expressed willingness to participate in Fissile Material Cut-off Treaty (FMCT) enabling the limiting of the number of warheads. India has been observing voluntary moratorium of nuclear tests. In the area of Nuclear Non Proliferation, India's good conduct is well recognised by the US. Hence, US has agreed to enter into Civilian Nuclear Cooperation with India.

Economic Development and Technological Leapfrogging

India believed in the necessity of nuclear technology for economic development especially for developing countries that are endeavouring to leap across the technology gaps. This thinking was reflected in the enactment of the Atomic Energy Act of 1948, within a year of India's independence. To justify the peaceful use of nuclear weapons, India has constructed most of the reactors for electricity purposes. Out of the 25 existing nuclear reactors, 17 are for electricity purpose and eight are for military purpose. In the following subsections, we discuss about the evolution of India's peaceful nuclear programme with and without the cooperation from foreign countries in three phases.

First Phase

During the first phase of India's peaceful nuclear programme (between 1947 and 1974), India actively pursued collaboration with the western countries *viz.*, the USA, the Canada and France in the construction of nuclear reactors, maintenance of nuclear reactors, transfer of technology and the import of uranium.

As a result, two boiling water (BW) nuclear reactors of 160 MW each known as, Tarapur Atomic Power Station (TAPS-I) and Tarapur Atomic Power Station (TAPS-II) were set up with the US assistance in Tarapur (in the state of Maharashtra) and one pressurised heavy water (PHW) reactor with a capacity of 100 MW known as, Rajasthan Atomic Power Station (RAPS-I) was set up with Canadian assistance at Rawatbhata in the state of Rajasthan.

TAPS-I and TAPS-II were set up by Bechtel and General Electrical (GE) under the '123' Agreement of 1963 between India, US and IAEA. The reactors are functioning since October 1969. RAPS-I is functioning since

December 1973, about 55 per cent of the components in RAPS-I are indigenously manufactured. This apart, RAPS-II with a capacity of 200 MW was planned and Canadian assistance was forthcoming. (See wikipedia articles on Tarapur Atomic Power Station and Rajasthan Atomic Power Station.)

Second Phase

India's nuclear test conducted in September 1974 commenced the second phase of India's peaceful nuclear programme which continued until 2008 when the Indo-US nuclear deal was signed. The test was interpreted as hostile by Western countries including the US and Canada. This lead to considerable delay in the construction of RAPS-II.

Finally, RAPS-II commenced operation in October 1981 with about 75 per cent of components manufactured indigenously.

Apart from RAPS-II, 13 other nuclear reactors of varying dimensions were completed during the second phase. RAPS-II apart from 11 other reactors *viz.*, RAPS-III and IV, Kakrapar Atomic Power Station (KAPS-I and II), Madras Atomic Power Station (MAPS-I and II), Narora Atomic Power Station (NAPS-I and II) and Kaiga Atomic Power Station (KAPS-I, II and III) reactors are of Canada Deuterium Uranium (CANDU) design.

The CANDU design was retained thanks to the Atomic Energy of Canada Ltd. (AECL) providing technical and design information without charge to the Energy Commission of India. Not withstanding the international non cooperation, Indian nuclear scientists managed to indigenise the production of nuclear reactors although with considerable time lag and cost escalation gradually overcome through the 'learning curve.' This cost and time escalation problems have been overcome under the guidance of the Nuclear Power Corporation of India Ltd. (NPCIL) established in 1986.

Dimensions of International Non-Cooperation

Turning to international non-cooperation, we can categorise it into eight components based on the Indo-Canadian negotiations between 1974 and 1976.

First, Canadian or for that matter US or French supplied nuclear equipment and material should not be used in nuclear explosion.

Second, the nuclear technology supplied by these countries should not be used in nuclear explosion.

Third, nuclear technology suppliers from Canada should get approval from the departments of industry, trade and commerce while trading with India. Similar Government regulations apply to suppliers of US, France and other exporting countries.

Fourth, the nuclear technology acquired by India from Canada or any other exporting country can be transferred to a third country only with the consent of the exporting country.

Fifth, reprocessing of uranium can be carried out only with the consent of the exporting country.

Sixth, enrichment of uranium is restricted to 20 per cent by Canada and prohibited by the US.

Seventh, fall-back options: including outright ban on reprocessing and enrichment can be exercised by Canada and other exporting countries.

Eighth, protection must be accorded to each of the reactors constructed with the assistance of the exporting countries *viz.*, the Canada. India agreed to provide protection only to RAPS-I and RAPS-II and did not agree to comply with the rest of the conditions (see *http://www.ccnr.org/exports_3.html*)

Negative Implications of International Non-Cooperation

India's peaceful nuclear programme felt immediate adverse impact due to international non-cooperation.

Firstly, the construction of reactors which ought to have taken 5 to 7 years took over 15 years. RAPS-II took 13 and MAPS-I and II took 17 and 15 years respectively.

Secondly, there were several safety problems that came to the notice of the international scholars. RAPS-I and RAPS-II reactors were shut down for three years between 1994 and 1997 to facilitate complete retubing using the zerconium-niobium pressure tubes discarding the zircaloy pressure tubes used hitherto in the CANDU nuclear reactors. MAPS-I and II reactors had serious problems with their heat transport system leading to downsizing of the capacity to 170 MW.

NAPS reactors suffered from serious accident in 1993 leading to the breaking of turbine blades, the consequent bursting of hydrogen and the destruction of turbine hall leading to the loss of coolants. Emergency coolant was used to maintain cooling. (See *http://www.ccnr.org/exports_3.html*).

KAPS-I suffered from the leakage of coolants and the reactors were shut down for 66 days in 1998. In 2004, KAPS-I's control rods were damaged during maintenance and poisons were added to the system.

Positive Implications of International Non-Cooperation

Indigeneous production of nuclear reactors enabled Indian scientists to learn the nuances of modernity, safety and life extension. This learning led to gradual reduction of time lag and cost in the construction of nuclear reactors. MAPS-I and II reactors took much time to complete as noted above and had to operate at a much lower plant load factor (PLF). While the reactor PLF is 220 MW, Indian scientists could achieve a PLF of 170 MW *viz.*, 78 per cent. It should be noted that MAPS reactors are the first indigenously built reactors.

Turning to foreign supplied reactors *viz.*, RAPS-I and TAPS-I and II, the PLF at the outset was 41 per cent (100 MW of 220 MW), 76 per cent each (160 MW each of 210 MW each) respectively. This implies that even at the outset, the PLF of the initially indigenously built reactors *viz.*, MAPS-I and II is superior to that of the foreign supplied reactors thanks to the perseverance of Indian scientists.

RAPS-II for which Canadian assistance was withdrawn due to India's 1974 nuclear test could achieve a PLF of 91 per cent *viz.*, 200 MW of 220 MW. (See *http://www.ccnr.org/exports_3.html*).

The problem of time escalation was gradually overcome through the construction of KAPS, NAPS and Kaiga reactors. KAPS reactors were built between 1984 and 1995 *viz.*, in nine and 11 years respectively which is sharply less than the 17 and 15 years taken for MAPS-I and II reactors. The first Kaiga power reactor was constructed in 11 years between 1989 and 2000. RAPS-III and IV reactors were built in less than nine years.

The cost escalation problems were not yet overcome. The KAPS reactors were constructed with very high cost. Against the estimated cost of

Rs 3.8225 billion, the actual cost was Rs 13.35 billion. It should be noted that the construction of KAPS reactors commenced prior to the establishment of NPCIL and completed much after the establishment of NPCIL. Cost reduction was gradually visible through the construction of RAPS-III and IV reactors which were constructed with Rs 2,450 crore.

Compared to KAPS-I whose construction cost was Rs 1,335 crore *viz.*, Rs 6.135 crore per MW during late 1980s and early 1990s, the per capita cost of RAPS-I and II was Rs 1,225 crore *viz.*, Rs 5.57 crore per MW. There was a marked reduction in the time and cost in the construction of TAPS-III and IV reactors which commenced functioning in 2005 and 2006 respectively. RAPS-III and IV reactors were constructed in 6.5 years *viz.*, seven months ahead of the schedule.

The MW per cost was Rs 5.4 crore viz., Rs 5,832 crore on the whole against the estimated cost of about Rs 6,600 crore.[1] The TAPS-III and IV reactors have achieved the PLF of 92 per cent *viz.*, 500 MW each of the estimated 540 MW each.

Third Phase

The third phase in India's nuclear programme begun with the Indo-US nuclear deal coming into effect from September 2008. Prior to the negotiation on the deal, India has proven track record in nuclear non-proliferation. After the expiry of the agreement with US concerning restrictions on the proliferation pertaining to TAPS-I and II reactors, India has volunteered to place TAPS-I and II under the IAEA safeguards.

In 2005, India enacted the Weapons of Mass Destruction (WMD) law to prevent proliferation of and brokering on sensitive technologies. (See *http://www.cfr.org/publication/9663/*).

Even before the Indo-US nuclear deal was envisaged, small scale assistance was received by India from France, Canada, China and Russia. After the US stopped the supply of uranium to TAPS reactors in 1983, France supplied between 1983 and 1993, China between 1995 and 2003 and Russia from then on. Canadian AECL (Atomic Energy Canada Ltd.)

1. Paper by Dr. Nagaich presented at the international conference, *India and Russia: Problems in Ensuring Energy Security*.

assisted in safety and quality assurance by helping in the retubing of RAPS-I and II during 1994-1997. (see *http://www.ccnr.org/exports_3.html, http://www.cfr.org/publication/9663/ and http: //www.Hinduonnet.com/fline/fl2307/stories/20060421006811800.htm*).

The Indo-US Nuclear Deal

The US congress passed the 123 or the Indo-US nuclear deal in October 2008 beginning a new era in Indo-Us civilian nuclear cooperation and cooperation in satellite technology.

Conditions of the Deal

According to the deal, India agreed to place 14 out of 22 reactors under the IAEA inspection on a permanent basis. Most of these reactors are domestically built during the second phase of India's nuclear programme. Military facilities and the current stock piles of fissile materials are exempt from the inspections with India determining the right to classify military and civilian reactors.

India agreed upon continuing moratorium on nuclear testing, firmly securing the nuclear arsenal, entering into negotiations towards signing Fissile Material Cut-off Treaty (FMCT) and preventing the proliferation of reprocessing and enrichment technologies. In return for fulfilling these conditions, India will benefit from nuclear fuel and technology from US companies, direct US investment in India's nuclear industry and the supply of fuel and technology from the NSG countries.

Benefits from the Nuclear Deal

India can buy US dual-use nuclear technology, (including materials and equipment enabling the enrichment of uranium or reprocessing plutonium.

India can receive imported fuel for its nuclear reactors. (See *http://www.cfr.org/publication/9663/*).

The deal paves the way for India acquiring and building the Generation-III turnkey reactors: such as: ACR-1000 (the Canadian 1,000 MW Generation-III CANDU reactors), the AP-1000 (US Generation-III 1,000 MW Westing House Corporation designed reactors), the VVER-1000

(Rosatom—Russian company designed 1,000 MW Generation-III reactors) etc. However, the deal may result in capping the R&D on thorium-based Stage-III nuclear reactors.

It should be noted that India has the third largest thorium reserves in the world and the largest R&D programme on thorium-based reactors. This can enable India to be the energy super power towards the end of this century which would cater to the energy security requirements of the world. However, India's increased capability to participate in upsising the reactors in terms of MW through future generation reactors thanks to the Indo-US nuclear deal is in itself a welcome benefit for India. (See *http://www.dailykos.com/story/2008/9/9/192244/9707/224/592681*).

We should hope that the thorium programme will not all together be scrapped. Through international participation, we can hope to revive and fructify the thorium programme before the uranium reserves run out.

Nuclear Opportunities in India

This section discusses the opportunities bestowed by India's nuclear policy with special reference to the Indo-US nuclear deal in terms of energy security, economic benefits, technological cooperation and strategic benefits.

Energy Security

India is seriously short of electricity. The relative paucity of coal and natural gas and the environmental and technological benefits arising out of nuclear power makes nuclear power a better choice for India's energy security.

The following paragraphs discuss India's energy profile and the increasing potential of nuclear power in the wake of diminishing potential of fossil fuels *viz.*, oil, natural gas and coal.

India's Energy Profile

World's energy requirements are met by fossil fuels *viz.*, oil, coal and natural gas respectively. World's energy resources are finite. Oil is expected to run out in about 40 years, natural gas in about 60 years and coal in about 200 years respectively. With the current level of technology, renewable energy cannot be made the viable substitute to fossil fuel-based energy. Due to the paucity of endowments of water supply and finity of land, renewable energy is not an economic substitute for fossil fuel-based energy in the near future.

Biofuels have adverse impact on the availability of food grains causing heavy financial burden on the poor and imposes additional heavy burden of food subsidy on the governments and rationing on the people. Hence, Government should ensure energy security to India through a strategy recognising the indispensability of fossil fuel-based energy during the foreseeable future *viz.*, 30/40 years.

India is the 11th largest energy producer and sixth largest energy consumer in the world. Fossil fuels *viz.*, oil, coal and natural gas are important sources of primary energy consumption for India. Coal contributes to 54 per cent, oil to 34 per cent and natural gas, nuclear and other renewable energy sources to 12 per cent respectively. To a large extent, India depends on imports of fossil fuels (See USEIA, 2007).

Oil and Natural Gas

India has about 5.6 billion barrels of total proven oil reserves *viz.*, 2.4 per cent of the world. The probable and proven reserves would be up to 11 billion barrels. Most of India's oil is produced from onshore fields in Assam and offshore fields off the Mumbai coast. India is the 25th largest oil producer accounting for 1 per cent of global oil production and sixth largest consumer of oil accounting for 2.9 per cent of world's oil consumption.

India produces over 850,000 barrels of oil per day and consumes over 2,700,000 barrels of oil per day. India thus imports over 65 per cent of oil (see Appendix A-7.1 of this paper and USEIA, 2007).

The most important source of India's oil imports is the Middle East. Saudi Arabia is the largest oil exporter to India and Qatar and Oman are the largest natural gas exporters to India. India is attempting to diversify the sources of oil and natural gas imports by including Sudan, Nigeria, Syria, Vietnam, Russia, Myanmar etc. However, it is understood that this policy of diversification brings India into direct competition with China leading to rapid deterioration of non-Middle Eastern oil resources.

As a consequence, this would most likely make India even more dependent on Middle Eastern oil resources from the year 2020. This would no longer help India influence the price level of oil and natural gas and would subject India to scramble *vis-à-vis* the more confident and aggressive China. China has better bargaining power *vis-à-vis* India due to the ability to supply sophisticated arms and enter into technological cooperation in sensitive areas *viz.*, missile development, space application etc. (See *http://iags.org/n0121043.htm*)

Turning to natural gas, we understand that India holds the reserves of over 29.32 cubic feet of natural gas which is about 0.5 per cent of world's reserves. India consumes about 1.2 trillion cubic feet of natural gas per year. India imports about half of its natural gas requirements. The share of natural gas consumption is growing very fast *vis-à-vis* the growth of coal consumption.

US Energy Information Administration (USEIA) data reveals that India's coal consumption is growing at an annual rate of about 2.2 per cent while oil and natural gas consumption is growing at the rate of 4 per cent and 4.8 per cent respectively.

This percentage growth will decline for the coal and increase for natural gas. Natural gas, alongside oil is becoming the predominant source of total primary energy production. For details on consumption (TPEP) for India (See Appendix A-2.1 (chapter 2) and USEIA, 2007).

Coal

India is the third largest producer and consumer of coal after USA and China. India has about 10 per cent of total world's coal reserves. Most of the domestic reserves contain relatively high amount of ash and bitumen. Most of the coal mines are located in states of Bihar, West Bengal and Madhya Pradesh. Given the current level of production and consumption, India's coal reserves would last for more than 200 years.

India produces about 8.5 per cent of world's coal output and consumes about 9 per cent of world's coal output. About half of India's energy requirements are met by coal and 70 per cent of India's electricity is produced using coal. About 23 million tonnes *viz.*, about 5 per cent of

India's coal requirements are imported. Primarily, the imports are from Australia, New Zealand etc.

This is high quality *viz.*, coking coal being used in the steel industry. India only has 5 per cent of total coal reserves as coking coal (See USEIA, 2007 and *http://www.indiaenergyportal.org/overview_detail.php*).

Electricity

India is the sixth largest producer of electricity accounting for 4 per cent of production and 3.5 per cent of consumption. India's electricity production is likely to grow by 8 to 10 per cent per year in the decade of 2020-2030. (See *http://www.indiaenergyportal.org/overview_detail.php and http://www.cslforum.org/india.htm*)

Nuclear Power

India produces 3 per cent of its electricity using nuclear power *viz.*, 4,100 MW. Due to the paucity of uranium, the reactors until recently were able to produce electricity only up to 46 per cent capacity. Assured supply of uranium from the NSG countries can raise the capacity to 60 per cent and the share of nuclear power in total power to 15-20 per cent by the year 2020. To raise the share of nuclear power to 15-20 per cent, India has to increase the nuclear power capacity to 40,000 MW from the present 4,100 MW (See *http://knowledge.wharton.upenn.edu/india/article.cfm?articleid=4217* and *http://www.dae.gov.in/iaea/ak-paris0305.doc*).

Nuclear power has the potential to enable India to achieve complete energy independence. However, the prospect is distant. To achieve complete energy independence, Stage-III nuclear reactors have to be fully operationalised. First pressurised heavy water reactors stage use the finite uranium-235 and uranium-238 derived from uranium-232. U-238 helps the closing of nuclear fuel cycle multiplying the times of use of the same fuel.

Second stage fast breeder reactors uses the less finite plutonium-239 using uranium-238 as the blanket. This alone could more than tripple the total energy production of India to 420 gigawatts.

Third stage breeder reactors use thorium-232 potentially helping India's complete energy independence. India has world's largest thorium reserves (See *http://www.dae.gov.in/iaea/ak-paris0305.doc*).

The Indo-US Nuclear Deal and Ramifications

The details of the deal include the following:

- India agrees to allow inspectors from the International Atomic Energy Association (IAEA), the United Nations' nuclear watchdog group, access to its civilian nuclear programme. By March 2006, India promised to place 14 of its 22 power reactors under IAEA safeguards permanently.

Teresita C. Schaffer, Director of the South Asia programme at the Center for Strategic and International Studies (US), says these will include domestically built plants, which India has not been willing to safeguard before now. India has promised that all future civilian thermal and breeder reactors shall be placed under IAEA safeguards permanently. However, the Indian prime minister says, 'New Delhi "retains the sole right to determine such reactors as civilian."' According to him: 'This means that India will not be constrained in any way in building future nuclear facilities, whether civilian or military, as per our national requirements.'[1] Military facilities—and stockpiles of nuclear fuel that India has produced up to now—will be exempt from inspections or safeguards.

- India commits to signing an Additional Protocol which allows more intrusive IAEA inspections of its civilian facilities.
- India agrees to continue its moratorium on nuclear weapons testing.
- India commits to strengthening the security of its nuclear arsenals.
- India works toward negotiating a Fissile Material Cut-off Treaty (FMCT) with the United States banning the production of fissile material for weapons purposes. India agrees to prevent the spread of enrichment and reprocessing technologies to states that don't possess them and to support international non-proliferation efforts.

1. See *http://www.indianembassy.org/prdetail943/—%09—suo-motu-statement-by-prime-minister-dr.-manmohan-singh-on-discussions-on-civil-nuclear-energy-cooperation-with-the-us%3A-implementation-of-india's-separation-plan*

- US companies will be allowed to build nuclear reactors in India and provide nuclear fuel for its civilian energy programme. (An approval by the Nuclear Suppliers Group lifting the ban on India has also cleared the way for other countries to make nuclear fuel and technology sales to India.)

What Kind of Technology would India Receive in Return?

India would be eligible to buy US dual-use nuclear technology, including materials and equipment that could be used to enrich uranium or reprocess plutonium, potentially creating the material for nuclear bombs. It would also receive imported fuel for its nuclear reactors.

What do Proponents Say about the Deal?

Proponents of the agreement argue it will bring India closer to the United States at a time when the two countries are forging a strategic relationship to pursue common interests in fighting terrorism, spreading democracy, and preventing the domination of Asia by a single power. Ashley Tellis of the Carnegie Endowment for International Peace (CEIP), who was intimately involved in negotiating the civil nuclear agreement with India as senior adviser to the US Undersecretary of State for Political Affairs, said in congressional testimony in 2005 that the deal recognises this growing relationship by engaging India, which has proven it is not a nuclear proliferation risk. Other experts say the deal lays out the requirements for India to be recognised as a responsible steward of nuclear power. 'This is part of a process of making India a more durable and reliable nuclear partner,' Schaffer said. Other experts say, the deal: Would encourage India to accept international safeguards on facilities it has not allowed to be inspected before. This is a major step, experts say, because the existing non-proliferation regime has failed either to force India to give up its nuclear weapons or make it accept international inspections and restrictions on its nuclear facilities. 'President Bush's bilateral deal correctly recognises that it is far better for the nonproliferation community if India works with it rather than against it,' writes Seema Gahlaut of the University of Georgia's Center for International Trade and Security in a CSIS Policy Brief. IAEA Director-General Mohamed ElBaradei has strongly endorsed the deal, calling it a pragmatic way to bring India into the non-proliferation community.

Recognises India's History of Imposing Voluntary Safeguards on its Nuclear Programme

Proponents of the deal say India has an excellent record of setting credible safeguards on its nuclear programme for the last 30 years. After the safeguards on the US-supplied Tarapur nuclear facility expired in 1993, for example, India voluntarily established a new agreement with the IAEA to continue the restrictions.

Recognises that India has a Good Record on Proliferation

Although it is not a signatory to the NPT, India has maintained strict controls on its nuclear technology and has not shared it with any other country. Proponents of the deal say this restraint shows that India, unlike its nuclear neighbour Pakistan, is committed to responsible nuclear stewardship and fighting proliferation. In May 2005 India passed a law, the WMD Act, which criminalises the trade and brokering of sensitive technology.

Rewards India's decision to adopt similar nuclear export standards as those imposed by the Nuclear Suppliers Group (NSG). India has thus far chosen to abide by the strict export controls on nuclear technology set by the NSG, a group of 45 nuclear-supplier states that coordinates controls of nuclear exports to non-nuclear-weapon states.

Economic Benefits from Nuclear Deals with Foreign Countries

In the aftermath of the Indo-US nuclear deal, over $150 billion worth of nuclear reactors and uranium are likely to be manufactured for the use in India in the next three decades *viz.*, between 2009 and 2040. India targets to produce over 60,000 MW of nuclear power per annum by 2040 and 40,000 MW by 2024. The proportion of nuclear power in total power production is likely to go up from 3 per cent to 10 per cent. A substantial portion of investment is likely to flow into India as a result of the Indo-US nuclear deal and the resulting international cooperation.

This is because, foreign deals with over $66 million should result in investment in India for the purpose of manufacturing. Bulk of the nuclear reactors is likely to come from the USA. This is likely to largely benefit companies like AREVA, WestingHouse and GE (General Electrical). A smaller proportion of benefits are likely to accrue to Russia's Rosatom. GE itself expects to construct 30 nuclear reactors by 2030. From the Indian point of view, bulks of the benefits out of the nuclear deal are likely to be enjoyed by NPCIL. This is because, as per the regulations, 51 per cent of the holdings in a nuclear power company should be held by the Government.

Atomic Energy Commission (AEC) Chairman Mr Anil Kakodkar stated (2005 lecture at IAEA) that the Atomic Energy Act will be suitably amended to promote private sector participation in the nuclear industry. Tata Power Corporation expressed its willingness to operate a nuclear power plant. Apart from the Tata group, Reliance Energy, the Essar Group and the GMR Group are most likely to own and operate nuclear power plants. Reliance has set up a 'New Power Initiative' including senior executives from NPCIL. The Tata Group has recruited people with nuclear domain expertise. US companies helping to set up these plants will be looking to work with Indian contractors according to Mr Shivesh Kannavi.

Several Indian companies have built up expertise in the nuclear power domain during the time when India was subject to international sanctions led by the US. Some of the Indian companies include:

Larsen & Toubro (L&T), Hindustan Construction Company (HCC) and Gammon India Ltd. in civil construction; L&T in reactors; Bharat Heavy Electricals Ltd. (BHEL) in boilers; KSB, Kirloskar Brothers, Mather & Platt Pumps Ltd., Jyoti Ltd. and Bharat Pumps & Compressors Ltd. in boiler feed pumps; Alfa Laval, GEI Hamon Industry Ltd., Maharashtra Seamless Ltd. and Ratnamani Metals & Tubes Ltd. in heat exchangers; Honeywell Automation India Ltd. (HAIL) in panels and Rolta India Ltd. in consulting and engineering services. Some industry watchers also include Walchandnagar Industries Ltd. (WIL), Godrej & Boyce, Bharat Heavy Plates & Vessels (BHPVL), the Hyderabad-based MTAR Technologies Pvt. Ltd. (which produces assemblies and precision components for use in space and nuclear applications) and Crompton Greaves.

HCC was the first Indian company to undertake civil engineering works for the pressurised heavy water reactors. 'HCC has constructed four out of the seven nuclear power plants in India. Four new plants are under construction, with HCC building two of them.' L&T finds opportunities to set up 'light water nuclear reactors of the boiling water type or the pressurised water type.'

The technology for such reactors, which need enriched uranium as fuel, is available with the US, France, Japan and Russia. L&T is equipped to manufacture the main reactor vessels as well as steam generators. Other critical equipment for such nuclear power plants require imported raw materials which will be sourced in with relative ease in the aftermath of the Indo-US nuclear deal enabling the speeding up of the nuclear deal.

Given the extensive domestic experience and cost advantages, companies like L&T plan to export nuclear reactor building skills and associated operation and maintenance services. India's Uranium Corporation of India Ltd. (UCIL) is allowed to explore for uranium in foreign countries. Private companies are allowed to explore for uranium in India. Data collection and analysis will be outsourced to private companies by the Department of Atomic Energy (DAE). (See *http://knowledge.wharton.upenn.edu/india/article.cfm?articleid=4217 and http://www.reuters.com/article/americasDealsNews/idUSTRE4917X720081002*).

Technological Cooperation

Indo-US nuclear deal enables the way for India a acquiring bigger nuclear reactors. The present 17 nuclear reactors produce over 4,000 MW of nuclear power. On an average, it is about 240 MW per reactor. In contrast, US out of its 100 reactors produce 1,000 MW per reactor on an average. The plant life and safety record of India's reactors is esteemed to be high by the Indian scientists. In recent periods, many US reactors were closing down 10 years ahead of the completion of their plant life. It implies that many reactors were deemed economically unfit to function beyond 30 years. This paves the way for the US using Indian expertise to extend the lifetime of their reactors to 60 years.

Strategic Benefit

India's portfolio of nuclear equipment and material suppliers is expanded by the Indo-US Nuclear deal. India's energy security can be better enhanced by having several countries and companies as the 'strategic' options. For example, Rosatom of Russia, Areva of France, Toshiba of Japan etc., are serious competitors with GE and Westing House Corporation of the US with regard to the manufacture and supply of Generation-III 1,000 MW reactors.

Challenges to India's Nuclear Programme

From the energy security angle, India faces several challenges in disguise. Firstly, foreign policy serves as as an effective disguise. India and China are seriously competing for oil and natural gas resources in Africa, the West Asia and in Southeast Asia including in Myanmar. To prevent India from actively cooperating with many other members of the NSG, US can coerce India to support its foreign policy stance. For example, US can coerce India to vote against Iran in the UN Security Council and General Assembly resolutions.

If voting against Iran or any other Middle Eastern countries, India's energy security concerns in the oil and natural gas fronts would be seriously harmed. India's energy security through nuclear power is a very distant prospect. Even 1/3rd of India's electricity requirements may not be produced through nuclear option by 2050.

The perception of India as a counterweight to China in Asian power politics by the US can result in the escalation of economic and military rivalry between India and China. This situation would cause India to have little option except to forge close ties with US. This can jeopardise the fast growing India-China trade. (See *http://intellibriefs.blogspot.com/2008/06/india-ull-text-of-nuclear-scientists.html and http://www.indiadefence.com/analysis.doc*).

Secondly, geopolitical challenge is an important issue India has to reckon with. It is well known that India imports 70 per cent of oil—much of it from the Middle East and the Persian Gulf which are geopolitically sensitive areas. It is important for India to have friendly relationship with the Oil Producing and Exporting Countries (OPEC) including those in the Middle East. In the contingency of the future Iran-Pakistan-India gas pipeline or the Trans-Afghan gas pipeline being subject to security threats, India should be able to bank on the support from friendly countries of the OPEC.

Thirdly, strategic challenge is an important issue. US without doubt has strategic objective in entering into the 123 Agreement with India. US can subject the non-US civilian nuclear reactors to much stringent and frequent inspection by strategically influencing IAEA. This would further reduce India's choice of international partners and would further force India to rely on US companies for nuclear technology. The stringent and frequent inspection of non-US reactors would cause a dent in India's quest for self-reliance.

This was implied in the letter to the Prime Minister of India by leading nuclear scientists led by the former Director of Atomic Energy Commission Mr Gopala Krishnan.

The letter also expressed concern over the possible delay or the shelving of research on thorium-based reactors. India should aspire to overcome the geopolitical and strategic challenges through trade and investment channels.

The comparative advantages of various nuclear reactor producing companies in their technological and economic aspects should be understood. US and India can enter into partnership by establishing a 'Special Purpose Company through the principles of joint responsibility and joint risk management.' The company should act to meet the energy requirements of vast Indian and US markets.

This special purpose company should conduct feasibility studies, place tenders for reactors and operate the reactors—including the Indian and other non-US reactors. The principles of joint responsibility and joint risk management can substantially de-link the special purpose company from the US foreign policy agenda.

Once the joint Indo-US experiment is successful, similar special purpose companies can be set up by India and US to benefit other countries aspiring for nuclear power including Malaysia by enabling the use of appropriate combination of Indian, US and other technologies in the nuclear programmes of other countries.

Concluding Remarks

India's peaceful nuclear programme has evolved through various difficulties. There is a transformation in the US mindset that hitherto considered nuclear cooperation with India is a risk in terms of arms race and nuclear proliferation. The present state of understanding reflects the 'win-win' situation through Indo-US cooperation. This is especially because Indian scientists are confident in the ability of Stage-III thorium-based fast breeder reactors fully meeting the energy needs of the world although after many decades from now.

The increased self-reliance and self-confidence of India's nuclear establishment enabled the US to understand the 'win-win' situation arising out of the Indo-US nuclear deal. Commercial principles should override the foreign policy agenda of US and through joint enterprise, US and India should utilise the best technologies of all the companies producing nuclear reactors and adopt best practices for operating nuclear reactors.

8 India's Transition to Renewable Energy

Introduction

Renewable energy comprises 10 per cent of India's electricity production and consumption and about 15 per cent of total primary energy consumption in India. Solar energy, wind power, biomass, hydroelectric power, tides and waves and geothermal energy are the most important forms of energy in India. Hydropower, solar energy and biomass are the main thrust areas for India to solve the energy problems. Hydropower is renewable, inexpensive, non-polluting and the current technology is well developed to realise its potential growth.

Relevance of Renewable Energy

Across the world, energy security, global warming concerns and stimulus packages for economic recovery would most likely drive cleantech investments on a faster pace. Several companies are capitalising on the green technology potential to comply with stricter carbon emissions regulations. In fact, green technology is being viewed as an investment opportunity which cuts down the operating costs and also bestows carbon credits.

Carbon emissions must begin to decline from around 2015 to prevent adverse impact of climate change as noted by Fourth Assessment Report of Intergovernmental Panel on Climate Change (IPCC). Annual global investments in renewable energy, energy efficiency and carbon capture and storage (CCS) technologies are expected to reach half-a-trillion dollars by 2020 equivalent to 0.44 per cent of global gross domestic product (GDP). Solar energy is poised to make substantial contribution to the renewable energy portfolio in the next 10 years and to the energy portfolio at large in the next 20-40 years. Growth in solar energy requires right kind of tax incentives, right level of feed-in tariffs and appropriate restrictions on carbon emissions.

This chapter is divided into 14 sections.

Section II discusses India's need to reduce the carbon emission through the increased use of renewable energy. Sections III to VI discuss the National Action Plan on the Climate Change (NAPCC), India's steps towards shaping India as a renewable energy-based economy, the need for poverty alleviation and human development. Sections VII to X discuss India's GHG emissions, recorded events of climate change, climate change projections for the 21st century and the possible impacts of the climate change. Sections XI and XII discuss India's policy structure towards the greenhouse gas (GHG) mitigation and the envisaged clean development mechanism projects. Section XIII tracks the global R&D efforts and the flow of human resources pertaining to the development and use of the renewable energy technologies. Section XIV presents the concluding remarks.

Renewable Energy and Carbon Emission Reduction

Government of India has identified the need to actively participate in carbon emission reduction as part of its NAPCC. The present fossil fuel-based energy consumption add to the carbon emissions and in turn to global warming and to erosion of livelihoods due to reduced agricultural yield, severely degraded livestock and marine stock and the erosion of soil. Coupled with this phenomenon, the exhaustion of fossil fuels in this 21st century will deliver the 'ultimate livelihood shock' to the population.

In this scenario, people cannot use pump sets to draw water on account of almost exhausted groundwater and due to the severe scarcity of electricity. People cannot move from place to place leading to reverse urbanisation and isolation of communities. Hospitals and emergency services cannot function due to lack of fuel and electricity significantly increasing the rate of mortality and morbidity.

To cut it short, the world's economy, educational and institutional capabilities and governance structures will be pushed back by hundreds of years. To make matters worse, severe deforestation and the neglect of social forestry will cause great scarcity for fuel wood itself. Under this scenario, people will either have to perish through hunger, the attack of wild animals or due to internal conflicts, or to survive through rude savage means of brute strength, adoption to raw meat and even through cannibalism.

Sustained availability of energy, sustainable lifestyle and the design of sustainable habitat are the most pressing requirements of the society which will prevent the society from falling into the gloomy trap. India's NAPCC is a significant step towards a sustainable and green future. Green energy, sustainable habitat through energy efficiency and waste recycling and converting waste to energy are among the corner stones of NAPCC.

National Action Plan on Climate Change (NAPCC)

India is faced with the challenge of sustaining its rapid economic growth while dealing with the global threat of climate change. This threat emanates from accumulated GHG emissions in the atmosphere that are anthropogenically generated through long-term and energy intensive industrial growth and high consumption lifestyles in developed countries.

With the increasing rate of economic growth, developing countries are also substantially contributing to the rise in GHG emissions. Besides, the impact of GHGs does not recognise local, international, intercontinental or even atmospheric boundaries. In other words, the harmful impacts spread across the boundaries. Hence, every country needs to make planned efforts and mitigate GHG emissions to enhance positive global environmental benefits.

India's NAPCC identifies measures that promote development objectives while enhancing the environmental benefits which in turn help mitigate the adverse impact of climate change. The action plan outlines a number of steps towards advancing India's development and climate change-related objectives of adaptation and mitigation.

Agenda Setting Towards the Lion's Share of Renewable Energy

The Fourth Assessment Report of the Intergovernmental Panel on Climate Change (IPCC-AR4) inferred from the data on temperature, sea level and snow cover in the northern hemisphere between year 1850 and present. The most important conclusion is that the 'warming of the earth's climate system is unequivocal'.

The global atmospheric concentration of carbon dioxide has increased from a pre-industrial (1850) measure of about 280 parts per million (ppm) to 379 ppm in 2005. Multi-model averages predict that the temperature increases during 2090-2099 *vis-à-vis* the 1980-1999 level may range from 1.1 to 6.4°C. The sea level would rise to the extent of 0.18 to 0.59 metres.

This phenomenon could lead to adverse impacts on freshwater availability, increased oceanic acidification, severely curtailed food production, severe flooding of coastal areas and the increased burden of vector borne and water borne diseases associated with extreme weather events.

Necessity of Poverty Alleviation

Economic reforms aim at faster rate of growth of the Indian economy. GDP growth rates between 2004 and 2008 averaged about 8 per cent. However, 27.5 per cent of India's population still lived below the poverty line in 2004-05 and 44 per cent of Indians are still without access to electricity. The Integrated Energy Policy of India and Approach Paper to the Eleventh Plan emphasise the primacy of rapid economic growth that enables poverty reduction.

The poor are the most vulnerable to climate change. The former Prime Minister, late Smt Indira Gandhi, had stated: 'poverty is the worst polluter'. Therefore, development and poverty eradication are essential drivers of adaptation to climate change. Climate change will have severe adverse impact on women. With climate change, there would be increasing scarcity of water, reduction in yields of forest biomass, and increased risks to human health making the children, women and the elderly most vulnerable: especially in a household setting. The possibility of decline in the food grains availability augments the threat of malnutrition. The cumulative effect on women would be catastrophic. To prevent this unforeseen and uncalled for scenario, the adaptation programmes should mitigate gender disparities.

Relationship between Human Development Index and Energy Consumption

The strong positive correlation between energy use and human development is well recognised. India needs to substantially increase its per capita energy consumption to provide a minimally acceptable level of well being to its people.

Carbon Dioxide Emissions in India

India's per capita CO_2 emissions are well below the world's average. Here are the numbers for the year 2004.

Table 8.1

A Comparison of India's Per Capita GHG Emissions with Major Countries and with the World Average

Country	*Per Capita Carbon Dioxide Emissions (Metric Tonnes)*
USA	20.01
EU	9.40
Japan	9.87
China	3.60
Russia	11.71
India	1.02
World average	4.25

Source: *http://www.india.gov.in/allimpfrms/alldocs/15651.doc*

India has a well-developed policy, legislative, regulatory and programmatic regime for promotion of energy efficiency, renewable energy, nuclear power, fuel switching, energy pricing reform, and for addressing GHG emissions in the energy sector. Consequently, the energy intensity of India's economy has come down sharply since the 1980s and compares favourably with the least energy intensive developed countries

Recorded Changes in Climate and Weather Events in India

India's Initial National Communication, 2004 (NATCOM 1) to United Nations Framework Convention on Climate Change (UNFCCC) has consolidated some of the figures.

Surface Temperature

At the national level, an increase of -0.4°C in surface air temperatures was recorded during the 20th century. The warming trend was recorded along the west coast, in central India, in the interior peninsula, and in northeastern India. However, cooling trends are recorded in north-west India and in parts of south India.

Rainfall

The national level rainfall pattern has not shown any significant change. However at the regional level, variations have been recorded. A trend of increasing monsoon seasonal rainfall was recorded along the west coast, in northern Andhra Pradesh, and in northwestern India to the extent of +10 per cent to +12 per cent of the normal level during the 20th century.

A trend of decreasing monsoon seasonal rainfall was recorded over eastern Madhya Pradesh, northeastern India, and some parts of Gujarat and Kerala (to the extent of -6 per cent to -8 per cent of the normal level during the 20th century.

Extreme Weather Events

Instrument records over the past 130 years do not indicate any marked long-term trend in the frequencies of large-scale droughts and floods. However in multi-decadal periods, more frequent droughts were followed by less severe droughts. There has been an increasing trend in severe storm incidence along the coast at the rate of 0.011 events per year. The states of West Bengal and Gujarat have reported such increasing trends.

Declining trend was recorded in Orissa. Goswami *et al.* (2006), analysed the daily rainfall data set and recorded a rising trend in the frequency of heavy rain events and a significant decrease in the frequency of moderate events over central India between 1951 and 2000.

Rise in Sea Level

Using the records of coastal tide gauges in the north Indian Ocean pertaining to 1960-2000, Unnikrishnan and Shankar (2007) estimated the annual sea level rise. The increase was to the extent of 1.06-1.75 mm per year. These rates are consistent with the annual global sea level rise of 1-2 mm estimated by the IPCC.

Impact on Himalayan Glaciers

The Himalayas are the home to one of the largest resources of snow and ice and its glaciers form a source of water for the perennial rivers such as the Indus, the Ganga and the Brahmaputra. These three rivers as we know are the lifeline for the majority of India's population. Glacial melt may impact the long-term lean-season flows of these rivers, causing adverse impacts on the economy due to inadequate availability of water and hydro-power.

The available monitoring data on Himalayan glaciers records the recession of some glaciers from the later part of the 20th century. However, the trend is not consistent across the entire mountain chain. It is too early to establish conclusions over the long-term trends, or their causation. The NAPCC is mandated with the collection, augmentation and refinement of the data on a persistent basis.

Some Projections of Climate Change in India for the 21st Century

The impact is derived from the anthropogenic emissions of GHGs.

a. Annual mean surface temperature would increase by 3 to 5°C under A2 scenario and by 2.5 to 4°C under B2 scenario of IPCC.

b. The warming would be more pronounced in the northern parts of India according to simulations by Indian Institute of Tropical Meteorology (IITM), Pune.

The IITM predicts an increasing summer monsoon intensity from 2040 to the extent of 10 per cent by 2100. IITM also predicts an increasing frequency and magnitude of extreme weather phenomenon.

Possible Impacts of Projected Climate Change

Impact on Water Resources

Changes in key climate variables, namely temperature, precipitation and humidity may have significant long-term implications on the quality and quantity of water. The Brahmaputra, the Ganga, and the Indus river systems would benefit from the melting snow in the lean season. But would be adversely affected by the decrease in snow cover.

A decline in total run-off is predicted for all the river basins except for Narmada and Tapti. For the Sabarmati and Luni basins, a decline of $2/3^{rd}$ is predicted. The freshwater sources near the coastal regions will suffer salt intrusion due to sea level rise.

Impact on Agriculture and Food Production

Food production in India is sensitive to variability in monsoon rainfall and intra-seasonal temperature changes. Studies by Indian Agricultural Research Institute (IARI) and others predict a greater loss in the *rabi* crop. Every 1°C rise in temperature reduces wheat production by 4-5 million tonnes. Small changes in temperature and rainfall have significant effects on the quality of fruits, vegetables, tea, coffee, aromatic and medicinal plants and *basmati* rice. Pathogens and insect populations are strongly dependent upon temperature and humidity, and changes in these parameters may change their population dynamics. Other negative impacts on agricultural and allied sectors include lower yields from dairy cattle and decline in the breeding, migration and the harvest of the fisheries.

Global reports indicate a loss of 10-40 per cent in crop production by 2100.

Impact on Health

Climate change may alter the distribution of important vector species like malarial mosquitoes and may increase the spread of the accompanying diseases to new areas. If there is an increase of 3.8°C in temperature and a 7 per cent increase in relative humidity the transmission windows (months during which mosquitoes are active) will be open for all 12 months in nine states in India. The transmission windows in Jammu and Kashmir and in Rajasthan may increase by 3-5 months. However, in Orissa and in some southern states, a further increase in temperature is likely to shorten the transmission window by about 2-3 months.

Impact on Forests

Hadley Centre for Climate Prediction & Research's regional climate model (HadRM3), the A2 and B2 scenarios and the biome vegetation response model are used by Ravindranath *et al.* (2006). They predict that 77 per cent and 68 per cent of the forest areas in the country are likely to experience shift in forest types. The forest produce and livelihood patterns would undergo a significant shift. Correspondingly, the associated biodiversity is likely to be adversely affected. NATCOM I projects an increase in the area under xeric scrublands and xeric woodlands in central India at the cost of dry savannah.

Vulnerability to Extreme Events

Heavily populated coastal areas are exposed to frequent adverse climatic events such as cyclones, floods and drought and large declines in sown areas is very likely in in arid and semi-arid zones. Large areas in Rajasthan, Andhra Pradesh, Gujarat and Maharashtra and comparatively small areas in Karnataka, Orissa, Madhya Pradesh, Tamil Nadu, Bihar, West Bengal and Uttar Pradesh experience frequent drought conditions.

About 40 million hectares of land is flood-prone. This includes most of the river basins in the north and the northeastern belt. This affects about 30 million people on an average each year.

Impact on Coastal Areas

A mean sea level rise (SLR) of 15-38 cm is projected along India's coast by the mid-21st century and of 46-59 cm by 2100. India's NATCOM I assessed the vulnerability of coastal districts based on physical exposure to SLR, social exposure based on the affected population, and economic impacts. In addition, a projected increase in the intensity of tropical cyclones poses a serious threat to the heavily populated coastal zones in the country.

India's Policy Structure Relevant to GHG Mitigation

India has in place a detailed policy, regulatory, and legislative structure that relates strongly to GHG mitigation: The Integrated Energy Policy was adopted in 2006. Some of its key provisions are:

- promotion of energy efficiency in all sectors,
- emphasis on mass transport,
- emphasis on renewables including biofuels plantations,
- accelerated development of nuclear and hydropower for clean energy, and
- focussed R&D on several clean energy-related technologies.

Several other provisions relate to reforming energy markets to ensure that energy markets are competitive, and energy prices reflect true resource costs. These include: Electricity Act, 2003, Tariff Policy, 2003; Petroleum & Natural Gas Regulatory Board Act, 2006, etc. The provisions taken together are designed to:

- Remove entry barriers and raise competition in exploration, extraction, conversion, transmission and distribution of primary and secondary energy.
- Accomplish price reform, through full competition at point of sale.
- Promote tax reform to promote optimal fuel choices.
- Augment and diversify energy options, sources and energy infrastructure.
- Provide feed-in tariffs for renewables (solar, wind, biomass cogeneration).
- Strengthen, and where applicable, introduce independent regulation.

The National Rural Electrification Policy, 2006, promotes renewable energy technologies in rural areas where grid connectivity is not possible or cost-effective. The New and Renewable Energy Policy, 2005, promotes utilisation of sustainable, renewable energy sources, and accelerated deployment of renewables through indigeneous design, development and manufacture.

CDM-based Development: Journey Towards a Renewable Energy Scenario

Renewable energy projects meet the clean development mechanism (CDM) criteria. The CDM literally means the zero GHG emission regime. However, for our practical purpose, low emissions also are considered as part of the CDM criteria. We hereby discuss the opportunities and priority in the form of the CDM projects.

CDM Project Opportunities

Usually, investors tend to prefer CDM projects satisfy three basic criteria. They are:

1. High possibility of government approval.
2. Low transaction costs.
3. Significant cost reduction and GHG reduction potential.

Energy Sector Priorities

The Government is actively trying to access green energy funds (GEF) to promote the CDM projects. Several types of CDM projects compliment the developmental objectives of the government. The Government has identified the following project ideas in accordance with its development objectives.

a. Removal of barriers to the implementation of biomass-based power generation projects.

b. Removal of the barriers to achieving energy efficiency in the industries.

c. Carbon reduction through grid-interactive solar photovoltaic (PV) power generation.

d. Enhancing energy conservation by upgrading the steel re-rolling mills.

e. Increased use of the solar PV diesel hybrid systems for decentralised village electrification.

f. Cogeneration in cement plants.

g. Reduction of carbon emissions in power plants.

h. Decentralised rural electrification through biomass and solar energy with private sector participation.

i. Switching to non-petroleum fuels, electricity, fuel cells and hybrid engines in the transport sector.

j. Demand-side management in basic material and construction industry.

k. Improving the efficiency of rural agriculture pump sets.

l. Improving energy conservation in paper industry.

As we understand from the above list, the priority options for the Government of India are the efficient technologies in power generation, centralised and decentralised renewable energy options and energy efficiency in steel and paper industries. The text submitted by India on the 'flexible mechanisms' explicitly disallows the forestry sector projects.

Preference will be given to projects in the renewable energy arena and those focussing on improving the energy efficiency. Projects which help in improving the quality of life of the very poor from the environmental standpoint and in the promotion of the state-of-the-art, environmentally sound technologies are preferred most.

Ranking of the CDM Options

The preferred CDM options are ranked by the policymakers based on the following criteria:

a. Cost of mitigation.

b. Feasibility of the option.

c. Environmental and other benefits.

d. Consonance with the overall development priorities of the country.

Priority CDM Projects

Based on the information given above, the following projects in India could be proposed for CDM.

1. Power Generation

i. Coal power plant using integrated gasification combined cycle (IGCC) technology.

ii. Coal power plant using pulverised fluidised bed combustion (PFBC) technology.

iii. Renovation and modernisation of power plants.

2. Renewable Sources of Energy

i. Grid connected wind-based power generations.

ii. Grid connected solar-thermal energy for power generation.

iii. Wind pumps for agriculture.

3. Industrial Efficiency

i. Direct reduction processes in iron and steel industry.

ii. Continuous pulp digesters in pulp and paper industry.

iii. Demand-side management: efficient motors.

Most Practical Means of Undertaking CDM

Coal is the most important source of electricity and energy for India and it is the best place to commence the CDM projects. India currently has the installed power generation capacity of about 167,000 MW. This falls short by 11-15 per cent of the demand. The demand is accelerating each year and it is currently about 7.5 per cent. The demand for power in India is expected to be over 400,000 MW by the year 2020, requiring over 500 billion dollars of investment in the power sector alone. Out of the total power generation capacity, about 66 per cent is based on coal.

The proven reserves of coal are about 70 billion tonnes and estimated reserves are about 200 billion tonnes. Hence, coal will continue to be the dominant fuel for power generation in India for many decades to come. To control pollution and march towards sustainable development, introduction of cleaner technologies for the utilisation of coal is imperative.

The proposed CDM options are the IGCC and the PFBC respectively. This apart, renovation and modernisation of the power plants will go a long way in controlling GHGs and pollution.

PFBC is a clean and efficient technology for coal-based power generation. In this technology, the conventional combustion chamber of the gas turbine is replaced by a pressurised fluidised bed combustor. The products of combustion pass through a hot gas cleanup system before entering the turbine. Thus, the amount of CO_2 emitted through combustion gets substantially mitigated.

The CO_2 reduction from a per unit of power supplied at bus bar using PFBC *vis-à-vis* the conventional technology is 0.18 kg. Replacing a conventional plant of 500 MW capacity with a PFBC plant will result in CO_2 emission reduction of 0.58 million tonnes a year. The annual national mitigation potential of CO_2 through this technology option is estimated to be 8.166 million tonnes.

Since the PFBC technology entails a high capital cost, the cost of electricity production is marginally higher than that of the conventional power plant. However, the savings in the coal compensates for the higher costs. The technology considerably lowers the emissions of SO and NO.

Thus, the technology has great commercial and environmental significance. Power plant using the IGCC reduces the emissions burden and improves the net efficiency of electricity generation through coal.

The combined cycle technology centres around two elements:

1. A gasification plant that converts the fuel into a combustible gas and purifies the gas.
2. A combined cycle power plant which produces synthetic gas that fuels a gas turbine whose hot exhaust gases are used to generate steam to drive a steam turbine.

The efficiency of the IGCC is the product of the gasification system efficiency and the combined cycle efficiency. The auxiliary power consumption in oxygen blown gasifier is much higher than air blown gasifier and other pulverised coal plants. It is approximately 20 per cent of the gross output. The air separation plant is very energy intensive because of the cryogenic process involved for the production of oxygen. However, the net efficiency of this technology is significantly higher because of the higher calorific value of the synthetic gas produced. For the second-generation technologies, net efficiencies are expected to be in the range of 45 per cent-50 per cent.

The present coal-fired thermal power plants are expected to have a net efficiency of 36 per cent. Whereas, the state-of-the-art IGCC plants have net efficiencies of 46 per cent. The above estimates imply that an IGCC plant will reduce the CO_2 emission by 0.25 per KWh. A 500 MW IGCC power plant *vis-à-vis* the conventional coal-fired plant will reduce the coal consumption by about 0.69 million tonnes.

The proposed IGCC power plant will reduce emissions of SO_2, NOx, and suspended particulate matter (SPM). It will also reduce solid waste disposal by nearly 70 per cent compared to the conventional coal-fired plants. An IGCC plant consumes 0.46 kg of coal for one unit of power generated. Whereas, the conventional plants consume about 0.60 kg of coal per kWh. This technology decreases the coal consumed and the cost of transporting the coal. Demonstration projects help this technology to create the distribution channels for manufacturing and engineering activities related to gasification technology, advanced gas turbines, and other IGCC/ power-related activities.

Renovation and Modernisation of Power Plants

The current technology can meet the demand for electricity. However, replacement of the technology or modernisation results in the high capital cost but helps realise the efficiency. The present conventional coal-fired plants in India operate below 30 per cent efficiency level. Increasing the efficiency could lead to huge amounts of reductions in the emission of CO_2 from this sector.

But, renovation and modernisation has been very slow in the country mainly due to the paucity of funds. The proposed project looks into the renovation and modernisation of existing old inefficient power plants in the country. This will result in reductions in the emissions of SO_2, NOx, and SPM. The project would reduce coal consumption which in turn would greatly reduce CO_2 emissions.

Research and Development Initiatives Towards Clean Technology and Renewable Energy Generation

Japan, Germany, USA and other European Union countries along with South Korea and China play a significant role in the R&D activities concerning the renewable energy. South Korea and China are the emerging giants in the arena. We first discuss the case of the USA and that of India to compare India's position and discuss India's roadmap in R&D towards the development of the renewable energy technologies. We then discuss the R&D efforts of various countries in terms of the patent activities and the general direction of the patenting activities.

The Case of the USA

The USA spends significant amount of R&D funds on R&D in the renewable energy sector. Between 1978 and 2007, the Department of Energy spent more on R&D in renewable energy, nuclear energy and on the energy storage and electrical systems than on the fossil fuels. Even when we compare the period between 1948 and 2007, this holds true, although the nuclear energy benefitted more than the renewable energy and the energy storage and electrical systems.

In this context, renewable energy implies the solar energy, the wind energy and the biofuels to a large extent and tidal and geothermal energy to some extent along with the hydrogen. The following tables show the cumulative achievements and the percentages over the period.

Table 8.2

DoE Energy Technology Cumulative Funding Totals (Billions of 2008 Dollars)

Period Technology	*FY1998-FY2007 (10 Years)*	*FY1978-FY2007 (30 Years)*	*FY1948-FY2007 (60 Years)*
Renewable energy	3.94	15.43	16.96
Energy efficiency	6.02	14.18	14.32
Fossil energy	5.36	24.22	39.60
Nuclear energy	6.41	38.62	85.01
Electric systems	0.93	2.85	3.02
Total	22.66	95.30	158.91

Sources: DoE Budget Authority History Table by Appropriation, May 2007; DoE Congressional Budget Requests (several years); DoE (Pacific Northwest Laboratory), *An Analysis of Federal Incentives Used to Stimulate Energy Production*, 1980. *Deflator Source*: *The Budget for Fiscal Year 2009*. Historical Tables, Table 10.1. Gross Domestic Product and Deflators Used in the Historical Tables, 1940-2013. pp. 194-195.

Table 8.3

DoE Energy Technology Share of Funding

(Per cent)

Period Technology	*FY1998-FY2007 (10 Years)*	*FY1978-FY2007 (30 Years)*	*FY1948-FY2007 (60 Years)*
Renewable energy	17.4	16.2	10.7
Energy efficiency	26.6	14.9	9.0
Fossil energy	23.7	25.4	24.9
Nuclear energy	28.3	40.5	53.5
Electric systems	4.1	3.0	1.9
Total	100.0	100.0	100.0

Note: Derived from previous table.

Sources: DoE Budget Authority History Table by Appropriation, May 2007; DoE Congressional Budget Requests (several years); DoE (Pacific Northwest Laboratory), *An Analysis of Federal Incentives Used to Stimulate Energy Production*, 1980; DoE Conservation and Renewable Energy Base Table. February 1990. *Deflator Source*: *The Budget for Fiscal Year 2009*. Historical Tables. Table 10.1. Gross Domestic Product and Deflators Used in the Historical Tables, 1940-2013. pp. 194-195.

Indian Case

In India, not much information is available pertaining to the R&D budget of the companies involved in the renewable energy sector. Even the scarce information available does not elucidate on the R&D activities such as the adaption or the technology development etc. Government has identified renewable energy and clean technology sectors as focus areas in the coming decades. The international pressure to mitigate the impact of the climate change and Government's development imperative necessitating the availability of sustainable and high quality power is the major driver of the R&D initiatives for the aforesaid purposes.

The National Action Plan on Climate Change (NAPCC) announced on 30 June 2008 outlines various strategies to meet the challenges of climate change and focusses on increasing the proportion of renewable energy sources in the fuel mix, promoting energy efficiency, conservation of national resources and increasing the carbon sink. The National Solar Mission, the National Biofuel Policy and other policy interventions are the basis for new thrust in the R&D initiatives.

The National Solar Mission shall be discussed in detail in Chapter 10. However, it is imperative to state that the necessity of constructing the grid connected capacity of over 20,000 MW by 2022 and reaching up to 1,00,000 MW by 2030 requires a significant thrust in the R&D activities. Similarly, the ambitious target of blending 20 per cent of transportation fuel with the biofuel (biodiesel and bioethanol) also requires significant R&D efforts towards technology development and towards the adaptive measures.

Both public and private investments are taking place in the renewable energy and cleantech arena. Private equity and the venture capital play their role in the R&D initiatives and in the establishment of the projects. The dominant companies attracting venture capital investments in the field of renewable energy include the Tata BP Solar and the Green Infra respectively for solar and for wind energy respectively.

At the national level, the cess on the domestically mined and imported coal of Rs 50 per tonne would most likely generate Rs 30 billion for the National Clean Energy Fund (NCEF).

Similarly, the weighted deduction on the expenditure incurred on in-house research and development (R&D) is enhanced from 150 per cent to 200 per cent. The weighted deduction on payments made to national laboratories, research associations, colleges, universities and other institutions engaged in scientific research is enhanced from 125 per cent to 175 per cent.

Indicator of the R&D Efficacy: The Patents

Patents are the indicator of the R&D efficacy pertaining to a nation, an industry and even to the world at large. Concerning the renewable energy technologies, the largest number of patents are filed in the respective patent offices of Japan, the USA, the Germany and off late in South Korea and in the People's Republic of China. European Union experiences a steady growth from the early 1990s. But, the growth rates in the USA and Japan are higher from the early part of the 21st century and are very high in South Korea and in China respectively. In general, the growth rate was 10 per cent between 1990 and 2005 and 25 per cent between 2001 and 2005.

Geographic and resource situations along with the R&D budgets and the supporting policies have enabled the growth of patent filing in the afore-discussed countries. While Japan, South Korea and China witnessed a predominance of domestic patent holders, the USA experienced an even distribution of the US nationals and foreigners among the patent holders.

Japan focussed primarily on the solar power, hydrogen and fuel cell technologies. USA focussed primarily on the bioenergy, geothermal energy, hydrogen and fuel cell technologies. European National Patent Offices witnessed the patenting based on the core strengths of their respective nationalities. The biggest patent office *viz.*, Germany focussed primarily on the solar and wind power technologies.

The Korean Intellectual Property Office (KIPO) of South Korea and the State Intellectual Property Office (SIPO) of China witnessed a very high growth rate in patenting as noted above. While South Korea focussed more on the wind power, hydrogen and fuel cell technologies, China focussed more on the solar power and hydropower technologies. On the whole, original technologies were developed more in Japan, Europe and in the USA as compared to the China and South Korea. Original patents imply the product patents. China and South Korea witnessed a very high growth rate of PCT based patents along with the USA.

Role of the Governments

The role of the governments in fostering the renewable energy technologies is significant. Japan focussed on direct funding along with the regulations and subsidies to the solar PV modules under its advanced PV power generation programme. USA and Brazil significantly subsidised their bioenergy programmes. Countries generally focus on regulations such as the mandatory share of the renewable energy in the total energy mix and its escalation, provision of the subsidies and in the creation of the specialised infrastructure such as the Solar City, Biotech Park, the Silicon Valley etc.

In general, there is a notable correlation between the government's R&D funding and an increase in the number of patents. This trend can be discerned between 1996 and 2005 in Japan, the USA and in Germany with high R&D funding and in Australia, Norway and Austria with low R&D funding. However, South Korea experienced a higher growth rate in patent filings than the Germany despite a low government R&D funding.

USA, Japan and Germany spend 15, 7 and 25 per cent of their total energy budgets on R&D respectively. Bulk of the budget in USA *viz.*, 60 per cent is for bioenergy, hydrogen and for fuel cells along with a 20 per cent share for solar energy off late while Germany provides the bulk of the budget for solar and wind power. Japan focusses more on hydrogen and fuel cells providing 70 per cent of the budget and solar by providing 25 per cent of the budget. From 2003, geothermal energy is taken off the support list.

South Korea fully funds the R&D activities of the universities and the research institutes while the companies are funded based on their size, the level of development of the concerned technology, the probability of commercial success and its potential of the deployment. Solar PV and wind power, hydrogen and fuel cell technologies are receiving the support.

In general, adaptive R&D initiatives are the major support focus in many countries. Technologies are dependent on the geographical terrain and the resource potential of a particular region. Identification of suitable technologies and processes is necessary before arriving at the decision for adopting the relevant technologies, scales and processes.

Sources for Technological Innovation

Citations in the patent applications reflect the sources of the technology *viz.*, whether the patent is for the basic science or for the applied technology. In terms of patent and non-patent citations, Switzerland, the United States and Finland appear to orient themselves mostly towards basic science, having a particularly high ratio of patent to non-patent citations. Basic science builds the foundations for technological innovation, and the research from universities and research institutes can be developed into commercial technologies, creating new markets and providing impetus for economic development.

Technology Focus

Geographical and resource position warrant the technology focus of a country.

At European Patent Office (EPO), Japanese and US applicants account for the largest number of alternative energy patent filings. While Japanese applications focus primarily on hydrogen and fuel cell technologies, US applications are distributed relatively equally between CCS, wind power, hydrogen and fuel cell, bioenergy and solar energy technologies.

Among the European countries which trail Japan and the United States in terms of number of patent applications, Germany, France, the United Kingdom, Italy and Switzerland are dominant. Applications from Germany concentrate particularly on CCS and those from Denmark and Spain on wind power respectively. Canadian applications focus mainly on hydrogen and fuel cell technologies. The distribution of German applications at EPO among different technologies were as follows between 1978-2005: 35 per cent were filed for CCS technologies, 12 per cent for solar energy, 12 per cent for bioenergy, 11 per cent for wind power and 11 per cent for waste to energy respectively. In accordance with the rapid increase in the market share, bulk of R&D budget was allotted to solar energy.

In France, 20 per cent of the filings were for hydrogen and fuel cells followed by 17 per cent for CCS, 16 per cent for wind power, 13 per cent for hydropower, 11 per cent each for wave and tidal power and 9 per cent for bioenergy respectively. In the UK, 24 per cent of the filings were for CCS, 13 per cent for hydropower, 12 per cent each for hydrogen and fuel cell and waste to energy, 11 per cent for bioenergy, 10 per cent for wind power and 9 per cent for solar energy respectively. In the European Union as a whole between 2000 and 2005, hydrogen and fuel cells received major proportion of the funding followed by the wind power with 23 per cent and solar power 16 per cent respectively. The renewable energy R&D budget was 16-fold in 2001-2005 compared to the 1997-2001 period respectively. Australian renewable energy budget focusses mostly on R&D in solar power, hydrogen and fuel cells and bioenergy.

The above discussion and the predominance of the technology focus suggest that the US, Japan and Germany are famous for R&D in solar power; Germany and Denmark for wind power; the US, Germany, Sweden and Switzerland for bioenergy and the US, Germany, Japan and Canada for hydrogen and fuel cells respectively. Worldwide: especially in Canada, the US and even in Japan, the importance given to the geothermal energy is gradually decreasing.

The importance of bioenergy is increasing rapidly to cater to the automobile transportation purpose and even the hydrogen and fuel cells are finding place in civilian applications such as the automotives, the refineries etc. Prior to the 1990s, the hydrogen and fuel cell technologies were exclusively reserved for space and military applications. Solar power was transformed into electricity through the solar thermal and solar photovoltaic technology. The photovoltaic and solar thermal technologies are being used on a larger scale from the 1990s. Prior to this, the solar power was primarily used for heating.

In the case of Japan, solar power, hydrogen and fuel cells are projected to solve the energy problems. Hence, most of the Japanese filing patents in the JPO tend to focus on these areas. Foreigners filing the patents in the Japan Patent Office (JPO) tend to focus more on their national interest and hence, a diversity of patents is reported. In the case of South Korea, the R&D initiative focusses more on the ultra thin film solar modules and the hydrogen and fuel cells that provide energy to the automobiles.

The Chinese patents in the SIPO focus more on the solar power and biofuels. The quest for the commercial use of the wind, solar, hydrogen and fuel cell technologies has prompted the Chinese government to foster renewable energy development. The Ministry of Science and Technology and the National Development and Reform Commission are responsible for providing human resources and coordinating the budget for research and development activities in renewable energy. The share of renewable energy in the total electricity use in China is 8 per cent in 2010 which would grow to 10 per cent and 15 per cent respectively by 2015 and 2020 respectively thanks to tax deductions, price support and direct investment from the part of the Government.

The distribution of patent applications at SIPO between different renewable energy technologies has been relatively equal and constant over time. The only exception is the hydropower-related applications, which have declined substantially.

Flows of Human Resources

Residence data for applicants and inventors derived from patent applications and the cited literature can provide information on the use of domestic human resources and the leakage. The ratio of domestically owned applications including those of the foreign inventors can be understood as the degree of influx of research capacity into a country, or brain gain. Conversely, the ratio of foreign owned applications including those of the domestic inventors can be seen as a measure of outflux of research capacity, or brain drain.

In terms of brain gain, Switzerland ranks first among the countries studied here, followed by Canada, the United Kingdom, France, Germany and Denmark. In terms of brain drain, the United Kingdom takes first place, followed by France, Switzerland, Israel, Canada and Germany. On the whole, European countries have a high level of both brain gain and brain drain. This suggests that the technology developed by the European inventors is exported internationally to a large extent. While brain gain and brain drain are balanced for Switzerland and Denmark, most countries including France, Germany, Israel and the United States have a higher level of brain drain than brain gain. This fact indicates that domestic technologies tend to be used to a great extent in foreign countries. Japan demonstrated low levels of both brain drain and brain gain. This indicates that the technology developed by the Japanese researchers is used domestically to a greater extent.

Countries citing the highest proportion of domestic applications compared to foreign applications are the United States, the United Kingdom, Israel, Switzerland, and the Netherlands. In contrast, Danish applicants primarily rely on prior French and German patent applications. The degree to which applications from a particular country's applicants are cited in subsequent applications indicates the degree of influence that country's technology exercises worldwide. US patent applications are particularly influential *vis-à-vis* other countries' patent applications. The United States and Japan have the highest proportion of domestic citations, which indicates a high degree of technology independence.

Concluding Remarks

The exhaustion of the fossil fuels and the GHG emissions from the fossil fuels warrant the development and use of the renewable energy technologies. India's NAPCC necessitates the reduction of GHGs and the promotion of the CDM projects. For this purpose, various programmes aimed at promoting eco-friendly and sustainable development is envisaged in the NAPCC.

Agriculture, renewable energy, green buildings, waste to energy, sustainable habitat etc., are the important components of the NAPCC. Important CDM projects concern with the cleaner use of coal and natural gas and the hybridisation of the technology. The PFBC and the IGCC are the important forms of CDM projects concerning the coal and natural gas. CDM projects can enhance the efficiency of the power plants by about 50 per cent. The National Solar Mission is part of the CDM mechanism.

Patents are the indicators of the depth and width of the technology development and absorption. The USA, Japan, Germany, UK, France and Canada lead the way in the development of original patents as well as the derived patents under the PCT regime. China and South Korea predominate in the PCT-based patents. Most countries mentioned above focus on the hydrogen and fuel cell research. The USA, Germany and Japan concentrate more on the solar power; Germany and the USA concentrate more on the wind power and China has special focus on developing the hydropower technologies.

Thus, the flow of human resources can be understood by referring to the patent citations. Most Western European Countries balance the brain drain and brain gain. The USA is susceptible to 'brain drain' while Japan is not inclined towards any of these tendencies.

9 Renewable Energy Programmes and Schemes in India

India has initiated various programmes and schemes for the advancement of renewable energy. The present growth momentum in the wind energy and the realisation of the massive potential in solar energy are the most important functions of India's Ministry of New and Renewable Energy (MNRE).

Having the goal of a near complete transition to renewable energy by 2050 or so, the utilisation of the biomass to its potential and the small hydropower potential also have their notable role to play. Other minor schemes pertaining to the solar and bioenergy such as the cookers, the biofuels are not covered in this edition of the book. However, they will find space in future as their role becomes more relevant and their employment and technology creation potential becomes more vivid.

The following Sections discuss about wind energy, solar energy, biomass energy and the small hydropower.

Wind Energy

The Background

Wind energy is one of the important sources of renewable energy in the world and more so for India. India ranks fifth in the generation of the wind energy after the USA, Germany, China and Spain. India has already harnessed 12,128 MW of wind power and has a total potential of over 48,000 MW with Karnataka, Gujarat and Andhra Pradesh in the forefront.

Table 9.1 and 9.2 list out the potential and the actual capacity achieved.

Table 9.1

Wind Power Potential in India

Sl No.	*State*	*Potential (in MW)*
1.	Andhra Pradesh	8968
2.	Gujarat	10645
3.	Karnataka	11531
4.	Kerala	1171
5.	Madhya Pradesh	1019
6.	Maharashtra	4584
7.	Orissa	255
8.	Rajasthan	4858
9.	Tamil Nadu	5530
	Total	48561

Source: Ministry of New and Renewable Energy. *http://www.mnre.gov.in/file-manager/UserFiles/wpp.htm*

Technology

The use of wind as a renewable energy involves the conversion of power contained in masses of moving air into rotating shaft power. The conversion process utilises aerodynamic forces (lift and/or drag) to produce a net positive turning moment on a shaft, resulting in the production of mechanical power which can be converted to electrical power.

The amount of wind power (P) harnessable at a particular site can be expressed as,

P = 1/2 r AV3 h Watts

where

P = The density of air, kg/m3

V = The wind speed in m/s

A = Swept area of the blades through which the wind blows, in m2

H = Efficiency of conversion (< 40%)

The wind speed therefore plays a vital role for energy generation as there is a non-liner (cubic) relationship between wind speed and available power.

The energy content in wind at different region varies with latitude, land-sea disposition, altitude and season. In India the factor which mostly

Table 9.2

State-wise and Year-wise Wind Power Installed Capacity (MW)
(upto 31.07.2010)

State	*Up to Mar' 2002*	*2002-03*	*2003-04*	*2004-05*	*2005-06*	*2006-07*	*2007-08*	*2008-09*	*2009-10*	*2010-11*	*2011-12*	*Total*
Andhra Pradesh	93.2	0.0	6.2	21.8	0.45	0.80	0.0	0.0	13.6	55.4	54.1	245.50
Gujarat	181.4	6.2	28.9	51.5	84.60	283.95	616.36	313.6	197.1	312.8	789.9	2966.30
Karnataka	69.3	55.6	84.9	201.5	143.80	265.95	190.30	316.0	145.4	254.1	206.7	1933.50
Kerala	2.0	0.0	0.0	0.0	0.0	0.0	8.50	16.5	0.8	7.4	0.0	35.1
Madhya Pradesh	23.2	0.0	0.0	6.3	11.40	16.40	130.39	25.1	16.6	46.5	100.5	376.40
Maharashtra	400.3	2.0	6.2	48.8	545.10	485.30	268.15	183.0	138.9	239.1	416.5	2733.30
Rajasthan	16.1	44.6	117.8	106.3	73.27	111.75	68.95	199.6	350.0	436.7	545.7	2070.70
Tamil Nadu	877.0	133.6	371.2	675.5	857.55	577.90	380.67	431.1	602.2	997.4	1083.5	6987.60
Others	3.2	0.0	0.0	0.0	0.0	0.0	0.0	0.0	0.0	0.0	0.0	3.20
Total	1665.7	242.0	615.2	1111.7	1716.17	1742.05	1663.32	1484.9	1564.6	2349.2	3196.7	17351.50

Source: Ministry of New and Renewable Energy. *http://www.mnre.gov.in/file-manager/UserFiles/wp_installed.htm*

governs the availability of wind energy at a particular site is its geographical location with respect to the monsoon wind. The availability of data on wind speed being a basic requirement for determining the feasibility of wind power generation at any site and due to the highly uneven distribution of wind speed over the country, an assessment of the wind resource over different regions was undertaken before any plans of harnessing the wind energy were drawn for implementation.

Demonstration Programme

An aggregate demonstration wind power capacity of 71 MW has been established at 33 locations in nine states *viz.*, Andhra Pradesh, Gujarat, Karnataka, Kerala, Madhya Pradesh, Maharashtra, Rajasthan, Tamil Nadu and West Bengal. Demonstration projects are implemented through the state governments, state nodal agencies (SNAs) or State Electricity Boards (SEBs). Demonstration projects will be taken up only in those states where commercial wind power projects have not yet been initiated/taken off.

Deployment

A total capacity of 12,128 MW has been established up to July 2010, mainly in Tamil Nadu, Gujarat, Maharashtra, Andhra Pradesh, Karnataka and Rajasthan. Wind electric generators of unit sizes between 225 kW and 2.1 MW have been deployed across the country. India now ranks 5th in the world after USA, Germany, China and Spain in grid connected wind power installations. A cumulative total of over 82 billion units of electricity have been fed to the State Electricity Grids.

Manufacturing Base of Wind Electric Generators

Wind electric generators are being manufactured in the country by a dozen manufacturers, through: (i) joint ventures under licensed production (ii) subsidiaries of foreign companies, under licensed production and (iii) Indian companies with their own technology. An indigenisation level up to 70 per cent has been achieved in machines of unit sizes up to 500 kW. The import content is somewhat higher in higher capacity machines. The current annual production capacity of domestic wind turbines is about 3,000 MW.

Booster Measures and Incentives

a. Guidelines for Setting up Projects

The Ministry has been issuing guidelines for wind power development since July 1995 in order to bring about balanced growth of the sector. These guidelines relate to preparation of detailed project reports, micro-sitting, selection of wind turbine equipment, operation & maintenance, performance evaluation, etc. A list of manufacturers of certified wind turbine machines is issued by C-WET on quarterly basis.

b. Promotional Policies

A package of fiscal and financial incentives is available which includes concessions such as 80 per cent accelerated depreciation, concessional custom duty on specified items, excise duty exemption, sales tax exemption, income tax exemption for 10 years, etc. In addition, State Electricity Regulatory Commissions (SERCs) are determining preferential tariffs. This apart, Government of India is providing the generation-based incentives (GBI) to the projects which could not avail the accelerated depreciation.

Indian Renewable Energy Development Agency Ltd. (IREDA) provides loan for setting up wind power projects.

c. Procurement Requirement

The borrower is required to follow transparent and competitive bidding procedure for procurement and shall demonstrate that the procurement procedures adopted by them are appropriate to the circumstances and that the quality goods, services and works are purchased at reasonable and competitive prices. The borrower shall provide all such information and documents reasonably required in connection with the procurement of any goods, services and works to be financed by IREDA. Wherever the loan is sanctioned against international lines of credit such as the World Bank, Asian Development Bank, KfW (Germany), etc., the relevant procurement procedures stipulated for competitive bidding process will have to be followed by the borrower.

Box 9.1

Incentives for the Wind Power Manufacturers

Central Incentives

A. Indirect Taxes

Description of Goods:

1. Wind operated electricity generators upto 30 kW and wind operated battery chargers upto 30 kW
2. Parts of wind operated electricity generators for manufacturers of wind operated electricity generators, namely

a. Special bearing	5%
b. Gear box	5%
c. Yaw components	5%
d. Wind turbine controllers	5%
e. Parts of the goods specified at (a) to (d) above	5%
f. Sensor	25%
g. Brake hydraulics	25%
h. Flexible coupling	25%
i. Brake calipers	25%

3. Blades for rotor of wind operated electricity generators for the manufacturers of wind operated electricity generators. 5%
4. Raw materials for manufacturer of blades for rotor of wind operated electricity generators 5%

Conditions:

A. If the importer at the time of importation furnishes in all cases, a certificate to the Deputy Commissioner of Customs or Assistant Commissioner of Customs as the case may be, from an officer not below the rank of Deputy Secretary to the Government of India in the Ministry of Non-Conventional Energy Sources recommending the grant of this exemption and in the case of the goods at (ii) to (v) the said officer certifies that the goods are required for the specified purposes.

contd...

...contd...

B. Furnishes an undertaking to the said Deputy Commissioner of Customs Assistant Commissioner to the effect that:

I. In the case of wind operated electricity generators upto 30 kW, or wind operated battery chargers upto 30 kW, he shall not sell or otherwise dispose off, in any manner, such generators or chargers for a period of two years from the date of importation in the case of wind operated electricity generators upto 30 kW, or wind operated battery chargers upto 30 kW, he shall not sell or otherwise dispose off, in any manner, such generators or chargers for a period of two years from the date of importation.

II. In case of other goods specified at (ii) to (v), he shall use them for the specified purpose and

III. In case he fails to comply with sub-conditions (i) or (ii), or both conditions, as the case may be, he shall pay an amount equal to the difference between the duty leviable on the imported goods but for the exemption under this notification and that already paid at the time of importation.

Devices/Systems Excepted from Excise Duty

1. Wind operated electricity generator, its components and parts thereof.
2. Water pumping wind mills, wind aero generators and battery chargers.

A. Sales Tax

Exemption/reduction in Central sales tax and general sales tax are available on sale of renewable energy equipment in various states.

B. Direct Tax

I. Eighty per cent accelerated depreciation on specified non-conventional renewable energy devices/systems (including wind power equipment) in the first year of installation of the projects.

II. Tax holiday on power projects.

Source: Excise Duty [Notification No.6/2002, dated 01/03/2002 (S.No.237 Non-Conventional Devices/Systems).

Centre for Wind Energy Technology (C-WET)

The Centre for Wind Energy Technology (C-WET), Chennai was established in Tamil Nadu in 1998 as an autonomous institution under the administrative control of the Ministry of New and Renewable Energy (MNRE). C-WET's main activities include resource assessment and testing & certification.

The below table shows the list of demonstration projects commissioned by the C-WET.

Table 9.3

State-wise Demonstration of Wind Power Projects

State	*Demonstration Projects (in MW)*
Andhra Pradesh	5.4
Gujarat	17.3
Karnataka	7.1
Kerala	2.0
Madhya Pradesh	0.6
Maharashtra	8.4
Rajasthan	6.4
Tamil Nadu	19.4
West Bengal	1.1
Others	3.3
Total	71.0

Solar Energy Programmes

The Background

India receives solar energy equivalent to over 5,000 trillion kWh per year. The daily average solar energy incident varies from 4-7 kWh per square metre depending upon the location. The annual average global solar radiation on horizontal surface, incident over India, is about 5.5 kWh per square metre per day. There are about 300 clear sunny days in most parts of the country. There are two ways of harnessing the solar energy *viz.*, the solar photovoltaic and the solar thermal, by direct conversion to electricity and heat energy respectively.

Currently Existing Solar Energy Programmes

As part of the renewable energy for urban, industrial and commercial applications, the solar energy programme and technology is being implemented by the Government of India. The programme comprises the following components.

a. Solar water heaters,
b. Green buildings,
c. Solar/green cities,
d. Solar photovoltaic systems,
e. Solar air heating/steam systems and
f. Akshay Urja shops.

Solar Water Heaters

The scheme on solar water heaters (SWHs) provides various incentives to the developers and users.

Salient features of revised scheme on SWHS:

- Provisions for users & motivators.
- Loans at 2 to 5 per cent to different categories of users. Interest free loans to domestic users of northeast, hilly states, islands and of chhattisgarh, Jharkhand and Uttarakhand. Thirty-one banks operating the scheme. Scheme to continue till 2009-10.
- Motivators to get Rs 200 per 100 litre per day (lpd) system from banks.
- Capital subsidy equivalent to upfront interest subsidy raised from Rs 1,100 to Rs 1,750 for institutions and from Rs 825 to Rs 1,400 for commercial establishments. Release through SNAs. In specific cases through municipal corporations (MCs). Service charges @ Rs 100 per sq m.
- Direct release to government departments/PSUs, if installed for own use. Includes Energy Service Company (ESCO) mode of implementation.
- Builders & developers/housing boards/development authorities of institutional/commercial establishments/RWAs to get similar subsidy. For housing complexes @ Rs 1,900/ sq m of collector area.

Provisions for banks/FIs:

- Interest subsidy and administrative charges to banks/FIs through IREDA.
- PSU banks allowed operating intermediary scheme at zero per cent interest rate as being done by IREDA. Manufacturers could participate. Banks to get administrative charges @ 1 per cent of loan disbursed.
- PSU banks allowed to obtain advance interest subsidy.

Provisions for manufacturers:

- Could become intermediary of PSU banks/IREDA. Support up to Rs 5 lakh/year for publicity to individual BIS/MNRE approved manufacturers on cost sharing basis through IREDA.
- Loans at 5 per cent to approved manufacturers for technology improvements and expansion of production facilities from IREDA as well as from PSU banks.
- Support for study tours/exposure visits abroad by industry delegation.

General provisions:

- Support up to Rs 10 lakh for publicity campaign to banks/FIs/intermediaries/SNAs/MCs/utilities. Higher monitoring charges to SNAs for domestic systems (Rs 100 per system).
- Support up to Rs 10 lakh to municipal corporations for making amendment in building bye-laws. Similar support also for announcing rebate in property tax and to utilities for announcing rebate in electricity tariff.
- Administration charges to SNAs for assisting MCs/utilities for announcing such rebates. Support up to Rs 2 lakh per activity for seminars/training/workshops/publicity etc., to banks/MCs/SNAs. Support for survey, evaluation study also.

Potential Electricity Savings through Solar Water Heating System (SWH)

The technical overall potential assuming that 75 per cent of *pucca* houses of the country occupied by the owners will have solar water heaters could be taken as 140 million sq m of collector area. The achievable/economic potential based on purchasing power of people/requirement of hot water in a year/availability of space for installation of system/availability of solar radiation etc., may, however, be taken as 35-40 million sq m of collector area.

Electricity/Diesel Savings

- A 100 lpd system (2 sq m of collector area) installed in a home can save 4-6 units of electricity/day depending on the place of installation and hot water use. On an average it could be taken as 5 units/day. Maximum average saving with 300 clear days, therefore, could be taken as 1,500 units/year.
- Assuming 300 days of solar hot water use in Bangalore and 150 days in Delhi, the savings could be 1,500 and 750 units per year respectively i.e., replacement of a 2 KW electric geyser working for 2 ½ hours in a day. Considering all parts of the country and maximum installations in areas where hot water requirement is more during the year, average saving could be taken as 1,200 units/year/100 lpd system.
- One million such systems installed will be able to save 1,200 million units of electricity/year.
- A 100 lpd system (2 sq m of collector area) installed in an industry can save around 140 litres of diesel in a year.

Peak Load Shaving

- One system of 100 lpd can replace an electric geyser of 2 KW capacity in a home.
- One million such systems will replace 1 million geysers of 2 KW capacity each in homes. Assuming that at least 50 per cent of geysers are switched on at a time, this will have a peak load shaving of 1,000 MW.

CO_2 Reduction

- A 100 lpd system on an average saves up to 1,500 units of electricity/year. To generate that much of electricity from a coal-based power plant, 1.5 tonne of CO_2/year is released in atmosphere.

> One million solar water heating systems installed in homes will, therefore, also result in reduction of 1.5 million tonne of CO_2 emission in atmosphere.

The cost indicated above includes the cost of collectors, insulated hot water storage tank, system piping and instrumentation, electrical back-up, controls, installation, and five-year warranty. Costs towards piping from system to utility points, heat exchangers (if any), transportation, may be extra. Cost over and above this amount shall be borne by the beneficiaries. Banks/FIs will not reject loans on account of higher cost of systems.

The solar water heating systems are progressing at a rapid pace of over 33 per cent per year since the past two years given the potential cost savings and financing options available to the customers.

The below table lists the year-wise achievements.

Table 9.4

Year-wise Achievement of Solar Water Heating Systems

Year	*Achievements (In sq m of Collector Area)*
Upto 2002-03	6,50,000
2002-03	1,00,000
2003-04	1,50,000
2004-05	2,00,000
2005-06	4,00,000
2006-07	4,00,000
2007-08	4,50,000
2008-09 (till 31.12.08)*	3,00,000
Total	26,50,000

Note: * Target : 6 lakh sq m. Target for Eleventh Plan: 5 million sq m of collector area. Until 31st March 2011, the total installed capacity was 45,00,000 SQM. The cumulative target for the years 2011-2012, 2012-2017, and 2017-2022 is 6 million, 1.2 million, and 20 Million SQM respectively.

Source: Ministry of New, and Renewable Energy. *http://www.mnre.gov.in/file-manager/.../Year_wise_installations_swhs.pdf*

By 31-3-2010, 3.5 million sq m of collector area has been achieved. However, this is far short of the China's achievement of over 125 million sq m as we will discuss in the next chapter which is dedicated to the international comparison of solar energy.

Successful Case Study on the Adaption of SWHs

Solar water heating systems are fast catching up in urban areas such as Bangalore, Pune, Hyderabad and Mysore. The programme of the Ministry of Non-Conventional Energy Sources (MNES) of soft loans and other incentives have created a favourable environment for use of solar water heating systems and a significant infrastructure for the manufacture, installation, financing and servicing has been developed. The low payback period for urban households is an added attraction. The result is that a number of builders and colony developers are now offering houses and apartment blocks fitted with solar water heating systems.

The latest to adopt this is Magarpatta City, a large housing complex which is coming up on the outskirts of Pune. Magarpatta is one of the biggest housing complexes in India covering over 550 acres. Magarpatta city is situated off Solapur Road at Hadapsar, Pune. Solar water heating systems are fitted as an amenity to its flat holders and as such its cost is included in the cost of the flat. The complex will eventually comprise about 10,000 dwelling units in the form of individual houses and apartment blocks of various sizes including some buildings with as many as 11 floors. The housing complex with each house having a solar water heater. The complex will also have several buildings that will accommodate information technology-related business. Currently about 3,194 solar collectors have been put in all the residential neighbourhood comprising about 3,500 flats in the Phase-I & II of Magarpatta City. The total capacity of the installed system is 4,03,150 lpd and the saving of carbon emission in tonnes per year is 6047.25. The electrical units saved per year are 68.94 lakh KwH.

Upon completion of Magarpatta City, the total solar collectors will be 7,160 in number. Approximately 9 lakh litres of water per day will be heated through these collectors. The savings in electrical units would be 14,548,000 units per year which would mean saving Rs 5.81 crore/year in cost of power. The savings in carbon emission would be 13,483 tonnes per year. When completed, the township will probably have the biggest concentration of solar water heaters in any residential complex in India. It should serve as a model for future townships and residential colonies in the country.

Green Buildings

To enable the energy saving architectural design and its implementation in the construction of the houses, the the Energy Research Institute (TERI) has developed Green Rating for Integrated Habitat Assessment (GRIHA) scheme. The scheme rates incentivises the energy saving constructions.

Here are the salient features.

- Revised scheme on 'Energy Efficient Solar/Green Buildings' (Issued vide sanction No. 3/5/2008-UICA (SE) dated 5th February 2009.

Keeping in view our climatic conditions, and in particular the construction of non-AC buildings, a national rating system—GRIHA for green buildings has been developed by MNRE in association with TERI. The rating system is suitable for all types of buildings in different climatic zones of the country and will be operated through GRIHA secretariat. It will operate on 100 point marking basis with 43 points related to energy and waste management and will be promoted through various incentives under the scheme as per below:

- Building owners,
- Reimbursement of 90 per cent of the registration-cum-rating fee for projects upto 5,000 sq m builtup area with minimum 3 star rating and for projects > 5,000 sq m builtup area with minimum 4 star rating.
- Architects/design consultants.
- Rs 2.5 lakh for projects upto 5,000 sq m builtup area with minimum 3 star rating and Rs 5 lakh for projects > 5,000 sq m builtup area with minimum 4 star rating.

Municipal Corporations/Urban Local Bodies

Rs 50 lakh to municipal corporations and Rs 25 lakh to other urban local bodies that announce rebate in property tax for green buildings and make it mandatory to get the new buildings under Government and public sector units rated under GRIHA. Scheme presently confined to commercial and institutional buildings including housing complexes with minimum built area of around 2,500 sq m. Release of incentives will be made by

MNRE on reimbursement basis through GRIHA secretariat after validation of star rating post-construction by the National Advisory Council.

Benefits of Green Buildings

- Reduced energy and water consumption without sacrificing comfort levels.
- Reduced destruction of natural areas, habitats and biodiversity, and reduced soil loss from erosion, etc.
- Reduced air and water pollution (with direct health benefits).
- Limited waste generation due to recycling and reuse.
- Reduced pollution loads.
- Increased user productivity.
- Enhanced image and marketability.
- No increase in building cost.

Annual Awards

Award of Rs 50 lakh to municipal corporation and Rs 25 lakh to other urban local body who performs best. Annual awards to 5 star rated buildings under GRIHA also.

Promotional Activities

Upto Rs 2 lakh for each activity to specialised institutions for organising workshops/seminars/training/publications/awareness campaigns etc. Separate support available for solar water heating and photovoltaic systems in green buildings. Details available at MNRE website (*www.mnre.gov.in*).

Solar Photovoltaic Systems

Most of the solar photovoltaic systems in urban areas are the rooftop systems. The grid connected systems shall be discussed in the next chapter.

Revised scheme on 'Demonstration and Promotion of Solar Photovoltaic Devices/Systems in Urban Areas & Industry' (Major focus on Rooftop SPV systems).

Salient Features and Financial Support

- Systems mainly for electricity conservation (target 750 kW during rest of Eleventh Plan),
- Solar street lights,
- Solar traffic signals,
- Solar blinkers,
- Solar power packs/inverters and
- Solar illuminating hoardings/bill boards.

Other systems of community use as felt necessary by implementing agencies at Rs 150 per watt of SPV panels upto a capacity of 1 kW each with required storage batteries (preferably 6 hours) to a maximum of 50 per cent of cost of system to urban local bodies/SNAs/institutions not availing depreciation benefits and at Rs 100 per watt to a maximum of 33 per cent of the cost of systems with similar conditions to commercial establishments/industry availing depreciation benefit. In specific cases, where battery storage is not required, the support will be at Rs 115 per watt and Rs 75 per watt respectively.

Maximum capacity of the systems to be supported in a state will not be more than 1/5th of the target for a particular year, out of which not more than 1/5th capacity will be supported in a single urban local body. Support for solar street lights will be limited to 50 per cent of total capacity of SPV systems/devices in an urban local body supported under the scheme. Implementing agencies will be free to decide about the capacity and other specifications of systems to be installed in their area.

Rooftop systems for diesel abatement (Target: 4.25 MW during rest of Eleventh Plan) (for diesel abatement in institutions, government buildings, commercial establishments e.g., malls, hotels, hospitals, nursing homes, industry & housing complexes facing huge power shortage during daytime) at Rs 75 per watt of SPV panels to a maximum of 30 per cent of the cost of systems to profit making bodies who can also avail accelerated depreciation benefits and at Rs 100 per watt to a maximum of 40 per cent of the cost of systems to non-profit making bodies. Systems could be with or without grid interactive. Support will be available for systems capacity varying between 25 to 100 kW. No restriction of targets to states. Proposals in prescribed format to be considered on first-cum-first basis through SNAs.

Benefits

- Will pay back the cost in 5 to 6 years.
- Life: 25 to 30 years.
- No air and noise pollution.
- No moving parts and zero maintenance except cleaning of SPV penal once in a while.
- Beneficiary to decide supplier on cost and quality competitive basis.
- A 100 kW capacity rooftop system could save around 50,000 litres of diesel per year.

  Note: Scheme being issued vide Sanction No. 3/7/2008-UICA(SE) dated 17th February 2009.

Biomass Power/Cogeneration Programme

The Background

Biomass has always been an important energy source for the country considering the benefits it offers. It is renewable, widely available, carbon-neutral and has the potential to provide significant employment in the rural areas. Biomass is also capable of providing firm energy. About 32 per cent of the total primary energy use in the country is still derived from biomass and more than 70 per cent of the country's population depends upon it for its energy needs. Ministry of New and Renewable Energy has realised the potential and role of biomass energy in the Indian context and hence has initiated a number of programmes for promotion of efficient technologies for its use in various sectors of the economy to ensure derivation of maximum benefits. Biomass power generation in India is an industry that attracts investments of over Rs 600 crore every year, generating more than 5,000 million units of electricity and yearly employment of more than 10 million man-days in the rural areas. For efficient utilisation of biomass, bagasse-based cogeneration in sugar mills and biomass power generation have been taken up under biomass power and cogeneration programme.

Biomass Power & Cogeneration Programme

The programme is implemented with the main objective of promoting technologies for optimum use of country's biomass resources for grid power generation. Biomass materials used for power generation include bagasse, rice husk, straw, cotton stalk, coconut shells, soya husk, de-oiled cakes, coffee waste, jute wastes, groundnut shells and saw dust etc.

Potential

The current availability of biomass in India is estimated at about 500 million metric tonnes per year. Studies sponsored by the Ministry has estimated surplus biomass availability at about 120–150 million metric tonnes per annum covering agricultural and forestry residues corresponding to a potential of about 18,000 MW. This apart, about 5,000 MW additional power could be generated through bagasse-based cogeneration in the country's 550 sugar mills, if these sugar mills were to adopt technically and economically optimal levels of cogeneration for extracting power from the bagasse produced by them.

Technology

Combustion

The thermochemical processes for conversion of biomass to useful products involve combustion, gasification or pyrolysis. The most commonly used route is combustion. The advantage is that the technology used is similar to that of a thermal plant based on coal, except for the boiler. The cycle used is the conventional ranking cycle with biomass being burnt in high pressure boiler to generate steam and operating a turbine with generated steam. The net power cycle efficiencies that can be achieved are about 23-25 per cent. The exhaust of the steam turbine can either be fully condensed to produce power, or used partly or fully for another useful heating activity. The latter mode is called cogeneration.

Cogeneration

In India, cogeneration route finds application mainly in industries. Sugar industry has been traditionally practicing cogeneration by using bagasse as a fuel. With the advancement in the technology for generation and utilisation of steam at high temperature and pressure, sugar industry can produce electricity and steam for their own requirements and surplus for the grid using the same quantity of the biogas. For example, if steam generation temperature/pressure is raised from 400°C/33 bar to 485°C/66 bar, more than 80 kWh of additional electricity can be produced for each tonne of cane crushed. The sale of surplus power generated through optimum cogeneration would help a sugar mill to improve its viability, apart from adding to the power generation capacity of the country.

Deployment

The Ministry has been implementing biomass power/cogeneration programme since mid-1990s. A total of 259 biomass power and cogeneration projects aggregating to 2,312 MW capacity have been installed in the country for feeding power to the grid. In addition, around 135 biomass power and cogeneration projects aggregating to 1,700 MW of electricity are under various stages of implementation. Cogeneration projects in sugar mills include 139 projects with installed capacity aggregating to 1,411 MW. Another 110 projects are under implementation aggregating to 1,440 MW. States which have taken leadership position in implementation of bagasse cogeneration projects are Andhra Pradesh, Tamil Nadu, Karnataka, Maharashtra and Uttar Pradesh. Also the leading states for biomass power projects are Andhra Pradesh, Chhattisgarh, Maharashtra, Madhya Pradesh, Gujarat and Tamil Nadu.

Manufacturing Base

Manufacturing capability exists in the country for the equipment/ machinery required for setting up biomass projects. Except for some critical control equipment, most of the equipments can be procured from indigeneous sources.

Boilers

A number of large manufacturers have established capabilities for manufacturing spreader stoker-fired boilers, travelling grate/dumping grate boilers; atmospheric pressure fluidised bed boilers and circulating fluidised bed boilers. Due to recent upsurge of interest in cogeneration for surplus power, leading manufacturers are further upgrading their capabilities for high efficiency boilers.

Steam Turbines

Almost all combinations—condensing, single extraction/double extraction condensing, back pressure, etc., are now being offered in the country with full after sales services guarantees. The efficiencies of turbines now being offered are comparable to the best in the world.

Other Equipment

Apart from the main equipment, there is a well established capability and capacity for manufacture of related equipment for use of biomass for energy including harvesters, balers, briquette equipment, handling and firing equipment, pollution control systems, etc., Many multinational companies have set up manufacturing facilities in the country for such equipment.

Promotional Policies

Besides the Central financial assistance (CFA), fiscal incentives such as 80 per cent accelerated depreciation, concessional import duty, excise duty, tax holiday for 10 years etc., are available for biomass power projects. The benefit of concessional custom duty and excise duty exemption are available on equipments required for initial setting up of biomass projects based on certification by Ministry. In addition, State Electricity Regulatory Commissions (SERCs) have determined preferential tariffs and renewable purchase standards (RPSs). IREDA provides loan for setting up biomass power and biogas cogeneration projects.

Table 9.5

Tariffs Announced by the State Electricity Regulatory Commissions: Regulatory Parameters for Biomass Power & Cogeneration Projects (as on 30.06.2010)

State	*Tariff Fixed by Commissions*	*Specified RPS (%)*
Andhra Pradesh	@ Rs 4.05/kWh, (2009-10) (BM)	Min. 5%
Chhattisgarh	@ Rs 3.90/unit, (2009-10) (BM)	5%
Gujarat	@ Rs 4.50/unit, (BM) @ 4.50/unit (Cogen) for first 10 yrs.	10%
Haryana	@ Rs 4.00/unit, BM, 3% escalation (base year 2007-08)	
	Rs 3.74/unit (Cogen)	10%
Karnataka	@ Rs 3.59/unit, (PPA signing date) Rs 4.14/unit (10th year) (Cogen) 4.13 (10th year) (BM) @ Rs 3.66 per unit (PPA signing date)	Min.10%
Kerala	@ Rs 2.80/unit, (BM) escalated at 5% for five years (2000-01)	3%
Maharashtra	@ Rs 4.79/unit (Comm yr.) (Cogen) (IT)	
	@ Rs 4.98 (IT) (BM) (for 2009-10)	6%
Madhya Pradesh	@ Rs 3.33 to 5.14/unit paise for 20 yrs. With escl. of 3-8paise	10%
Punjab	@ Rs 4.04/unit, (2009-10) BM escalated at 3% (Cogen) & 5% (BM)	
	@ Rs 3.81/unit (2009-10) Cogen	Min. 3%
Rajasthan	@ Rs 4.47/unit, (2009-10)-water cooled and Rs 4.96-air cooled condensers (BM)	7.5%
Tamil Nadu	@ Rs 4.50/unit, (2009-10) (BM)	
	@ Rs 4.37/unit, (2009-10) (Cogen)	Min. 13%
Uttarakhand	@ Rs 3.06/unit, (2010-11) (BM)	
	@ Rs 3.12/unit, (2010-11) (Cogen) (new projects)	9%
Uttar Pradesh	@ Rs 4.29/unit, for existing and 4.38 for new with escalated at 4 paise/year, base year (2006)	7.5%
West Bengal	Rs 4.36/unit fixed for 10 years (BM)	3.8%
Bihar	Rs 4.17/unit (2010-11)–(BM)	
	Rs 4.25/unit (2010-11)–existing (Cogen)	
	Rs 4.46/unit (2010-11)–new (Cogen)	1.5%

Note & Source: BM: Biomass; O.A.: Open access terms & conditions as for CERC and SERCs order (kindly contact concerned Regulatory Commissions for details).

Table 9.6

Central Financial Assistance (CFA) and Fiscal Incentives

CFA for biomass power project and bagasse cogeneration projects by private/joint/coop./public sector sugar mills

	Special Category States (NE Region, Sikkim, J&K, HP & Uttarakhand)	*Other States*
Project Type	*Capital Subsidy*	*Capital Subsidy*
Biomass Power projects	Rs 25 lakh X (C MW)^0.646	Rs 20 lakh X (C MW)^0.646
Bagasse cogeneration by private sugar mills	Rs 18 lakh X (C MW)^0.646	Rs 15 lakh X (C MW)^0.646
Bagasse cogeneration projects by cooperative/ public sector sugar mills		
40 bar & above	Rs 40 lakh *	Rs 40 lakh *
60 bar & above	Rs 50 lakh *	Rs 50 lakh *
80 bar & above	Rs 60 lakh *	Rs 60 lakh *

Note: * Per MW of surplus power@ (maximum support Rs 8.0 crore per project).

For new sugar mills, which are yet to start production and existing sugar mills employing back pressure route/seasonal/incidental cogeneration, which exports surplus power to the grid, subsidies shall be one-half of the level mentioned above.

@ Power generated in a sugar mill (-) power used for captive purpose i.e., net power fed to the grid during season by a sugar mill. Note: CFA and Fiscal Incentives are subject to change.

Source: Ministry of New and Renewable Energy. *http://www.mnre.gov.in/schemes/grid-connected/biomass-powercogen/*

CFA for bagasse cogeneration project in cooperative/public sector sugar mills implemented by IPPs/state government undertakings or state government joint venture company/special purpose vehicle (Urja Ankur Trust) through BOOT/BOLT model:

Project Type	*Minimum Configuration*	*Capital Subsidy*
Single cooperative mill through BOOT/Bolt model	60 bar & above	Rs 40 L/MW of surplus power*
	80 bar & above	Rs 50 L/MW of surplus power*

Note: Maximum support Rs 8.0 crore/sugar mill.

* Power generated in a sugar mill (-) power used for captive purpose i.e. net power fed to the grid during season by a sugar mill.

CFA for bagasse cogeneration project in existing cooperative sector sugar mills employing boiler modifications:

Project Type	*Minimum Configuration*	*Capital Subsidy*
Existing cooperative sugar mill	40 bar & above	Rs 20 L/MW of surplus power*
	60 bar & above	Rs 25 L/MW of surplus power*
	80 bar & above	Rs 30 L/MW of surplus power*

Note: Power generated in a sugar mill (-) power used for captive purpose i.e., net power fed to the grid during season by a sugar mill. CFA will be provided to the sugar mills who have not received CFA earlier from MNRE under any of its scheme.

CFA is subject to change.

Fiscal incentives for biomass power generation:

Item	Description
I. Accelerated Depreciation	80% depreciation in the first year can be claimed for the following equipment required for cogeneration systems: 1. Back pressure, pass-out, controlled extraction, extraction–cum-condensing turbine for co-generation with pressure boilers 2. Vapour absorption refrigeration systems 3. Organic ranking cycle power systems 4. Low inlet pressures small steam turbines
II. Income Tax Holiday	Ten years tax holidays.
III. Customs Duty	Concessional customs and excise duty exemption for machinery and components for initial setting up of biomass power projects.
IV. General Sales Tax	Exemption is available in certain states.

Small Hydropower Programme

The Background

Hydropower is a renewable, non-polluting and environmentally benign source of energy. It is perhaps the oldest renewable energy technique known to the mankind for mechanical energy conversion as well as electricity generation. Hydropower represents use of water resources towards inflation free energy due to absence of fuel cost with mature technology characterised by highest prime moving efficiency and spectacular operational flexibility. Out of the total power generation installed capacity of 1,48,265 MW (April 2009) in the country, hydropower contributes about 25 per cent i.e., 36,877 MW.

Table 9.7

State-wise/Year-wise List of Commissioned Biomass Power/Cogeneration Projects (as on 30.6.2010)

(In MW)

S.No	State	Upto 31.3.2003	2003-04	2004-05	2005-06	2006-07	2007-08	2008-09	2009-10	2010-11	Total
1.	Andhra Pradesh	160.05	37.70	69.50	12.00	22.00	33.00	9.00	20.00	—	363.25
2.	Bihar	—	—	—	—	—	—	—	—	9.50	9.50
3.	Chhattisgarh	11.00	—	—	16.50	85.80	33.50	9.80	43.80	—	199.90
4.	Gujarat	0.50	—	—	—	—	—	—	—	—	0.50
5.	Haryana	4.00	—	2.00	—	—	—	—	1.80	28.00	35.80
6.	Karnataka	109.38	26.00	16.60	72.50	29.80	8.00	31.90	42.00	—	336.18
7.	Madhya Pradesh	—	1.00	—	—	—	—	—	—	—	1.00
8.	Maharashtra	24.50	—	11.50	—	40.00	38.50	71.50	33.00	10.00	288.50
9.	Punjab	22.00	—	—	6.00	—	—	—	34.50	—	62.50
10.	Rajasthan	—	7.80	—	7.50	8.00	—	8.00	—	28.00	59.30
11.	Tamil Nadu	106.00	44.50	22.50	—	42.50	75.00	43.20	62.00	23.50	419.20
12.	Uttar Pradesh	46.50	12.50	14.00	48.50	—	79.00	172.00	194.50	14.00	581.00
13.	West Bengal	—	—	—	—	—	—	—	16.00	—	16.00
	Total	483.93	129.50	136.10	163.00	228.10	266.00	345.40	447.60	113.00	2372.63

Source: Energy Alternatives India. *http://www.eai.in/ref/ae/bio/csbg/statewise_installation.html*

Hydropower Project Classification

Hydropower projects are generally categorised in two segments i.e., small and large hydro. In India, hydro projects up to 25 MW station capacities have been categorised as small hydropower (SHP) projects. While Ministry of Power, Government of India is responsible for large hydro projects, the mandate for the subject small hydropower (up to 25 MW) is given to Ministry of New and Renewable Energy.

Small hydropower projects are further classified as:

Class	*Station Capacity in kW*
Micro hydro	Up to 100
Mini hydro	101 to 2,000
Small hydro	2,001 to 25,000

Small Hydropower Programme

SHP programme is one of the thrust areas of power generation from renewable sources in the Ministry of New and Renewable Energy. It has been recognised that small hydropower projects can play a critical role in improving the overall energy scenario of the country and in particular for remote and inaccessible areas. The Ministry is encouraging development of small hydro projects both in the public as well as private sector. Equal attention is being paid to grid-interactive and decentralised projects.

Aim: The Ministry's aim is that the SHP installed capacity should be about 7,000 MW by the end of Twelfth Plan. The focus of the SHP programme is to lower the cost of equipment, increase its reliability and set up projects in areas which give the maximum advantage in terms of capacity utilisation.

Potential: An estimated potential of about 15,000 MW of small hydropower projects exists in India. Ministry of New and Renewable Energy has created a database of potential sites of small hydro and 5,415 potential sites with an aggregate capacity of 14,305.47 MW for projects up to 25 MW capacity.

The following table shows the identified sites and their total capacity across the states.

Table 9.8

State-wise Details of the Identified/Future Sites

(State-wise Identified Small Hydel Sites and Potential) up to 25 MW Capacity (as on 31.3.2009)

S.No.	*Name of State*	*Identified Number of Sites*	*Total Capacity (in MW)*
1.	Andhra Pradesh	489	552.29
2.	Arunachal Pradesh	566	1333.04
3.	Assam	60	213.84
4.	Bihar	94	213.75
5.	Chhattisgarh	164	706.62
6.	Goa	9	9.10
7.	Gujarat	292	196.97
8.	Haryana	33	110.05
9.	Himachal Pradesh	547	2268.41
10.	Jammu & Kashmir	246	1411.72
11.	Jharkhand	103	208.95
12.	Karnataka	128	643.16
13.	Kerala	247	708.10
14.	Madhya Pradesh	99	400.58
15.	Maharashtra	253	762.58
16.	Manipur	113	109.10
17.	Meghalaya	102	229.81
18.	Mizoram	75	166.94
19.	Nagaland	99	196.98
20.	Orissa	222	295.47
21.	Punjab	234	390.02
22.	Rajasthan	67	63.17
23.	Sikkim	91	265.54
24.	Tamil Nadu	176	499.31
25.	Tripura	13	46.86
26.	Uttar Pradesh	220	292.16
27.	Uttarakhand	458	1609.25
28.	West Bengal	203	393.79
29.	A&N Islands	12	7.91
	Total	5,415	14,305.47

Source: EAI Whitepaper. *http://www.scribd.com/doc/37122894/EAI-White-Paper*. Accessed in February 2011.

Identification of new potential sites and strengthening of database for already identified sites is an ongoing process. In this direction, the Ministry has been giving financial support to state governments/agencies for identification of new potential SHP sites and preparation of a state perspective plan.

Small Hydro Installed Capacity and Progress

The total installed capacity of small hydropower projects (upto 25 MW) as on 31.03.2009 is 2429.77 MW from 674 projects and 188 projects with aggregate capacity of 483.13 MW are under construction.

Table 9.9

State-wise Numbers and Aggregate Capacity of SHP Projects (upto 25 MW) Installed & Under Implementation (As on 31.10.2011)

S. No.	State	Projects Installed		Projects under Implementation	
		Nos.	Capacity (MW)	Nos.	Capacity (MW)
1	Andhra Pradesh	64	192.63	18	62.05
2	Arunachal Pradesh	100	78.835	121	47.67
3	Assam	5	31.11	4	15
4	Bihar	19	59.8	9	24.1
5	Chattisgarh	6	19.05	7	148.2
6	Goa	1	0.05	-	-
7	Gujarat	5	15.6	-	
8	Haryana	7	70.1	2	3.4
9	Himachal Pradesh	126	440.965	33	132.25
10	J&K	35	130.59	5	6.65
11	Jharkhand	6	4.05	8	34.85
12	Karnataka	123	856.25	16	141.675
13	Kerala	21	141.67	13	60.75
14	Madhya Pradesh	11	86.16	3	4.9
15	Maharashtra	44	279.925	22	8.4
16	Manipur	8	5.45	3	2.75
17	Meghalaya	4	31.03	3	1.7
18	Mizoram	18	36.47	1	0.5
19	Nagaland	10	28.67	4	4.2
20	Orissa	9	64.3	4	3.6
21	Punjab	46	154.5	12	21.15
22	Rajasthan	10	23.85	-	-
23	Sikkim	17	52.11	1	0.2
24	Tamil Nadu	19	103.05	a	20.5
25	Tripura	3	16.01	-	-
26	Uttar Pradesh	9	25.1	-	-
27	Uttarakhand	97	146.82	50	217.25
28	West Bengal	23	98.4	17	84.25
29	A&N Islands	1	5.25	-	-
	Total	847	3197.795	356	1045.995

Source: Press Information Bureau, Government of India. *http://pib.nic.in/newsite/PrintRelease.aspx? relid=78107*

Privatisation Efforts

While in early 1990s, most of the SHP projects were set up in the public sector, from last 10 years or so, most of the capacity addition is made by the private sector projects. Beginning of the 21st century saw near commercialisation in the small hydro sector. Private sector entrepreneurs found attractive business opportunities in small hydro and state governments also felt that the private participation may be necessary in tapping the full potential of rivers and canals for power generation. The private sector has been attracted by these projects due to their small adoptable capacity matching with their captive requirements or even as affordable investment opportunities.

In line with Government of India policy, 18 states have announced their policy for inviting private sector to set up SHP projects. The Government of India announced the Electricity Act in 2003, Electricity Policy in 2005 and Electricity Tariff Policy in 2006 to create a conducive atmosphere for investments in the power sector. Small hydropower projects are now governed by these policies and the tariff is decided by the State Electricity Regulatory Commissions (SERCs) as per the tariff policy.

During the Tenth Plan, the following have been year-wise capacity addition from SHP projects.

Table 9.10

Year-wise Capacity Addition of the SHP Projects

Year	*Target (in MW)*	*Capacity Addition during the Year (in MW)*	*Cumulative SHP Installed Capacity (in MW)*
2002-03	80	80.39	1519.28
2003-04	80	84.04	1603.32
2004-05	100	102.31	1705.63
2005-06	130	120.80	1826.43
2006-07	160	149.16	1975.5
A target of adding 1,400 MW during the Eleventh Plan (2007-2012). During last two years, following have been the achievements:			
2007-08	200	205.25	2180.84
2008-09	250	248.93	2429.77

Source: Energy Alternatives India. *http://www.eai.in/ref/ae/bio/csbg/statewise_installation.html*

Standards for Small Hydro

In order to ensure project quality/performance, the MNRE has been insisting to adhere to International Electrotechnical Commission (IEC)/ international standards for equipments and civil works. The equipment in the project is required to confirm to the following IEC standards in order to avail subsidy from the MNRE.

Equipment standard turbines and generator (rotating electrical machines):

IEC 60034-1: 1983

IEC 61366-1: 1998

IEC 61116-1992

IS: 4722-2001

IS: 12800 (part 3) 1991

Field acceptance test for hydraulic performance of turbines:

IEC 60041: 1991

Governing system for hydraulic turbines:

IEC 60308

Transformers:

IS: 3156–1992

IS: 2705–1992

IS: 2026–1983

Inlet valves for hydropower stations & systems:

IS: 7326–1902

Scheme for SHP (upto 25 MW).

Recently the Ministry has given an assignment to Alternate Hydro Energy Center (AHEC), IIT Roorkee to revisit the existing standards and come out with standards/manuals/guidelines for improving reliability and quality of small hydropower projects in the country.

States with Policy for Private SHP Projects

Twenty-three states namely, Andhra Pradesh, Arunachal Pradesh, Assam, Bihar, Chhattisgarh, Gujarat, Haryana, Himachal Pradesh, Jammu and Kashmir, Karnataka, Kerala, Madhya Pradesh, Maharashtra, Manipur, Meghalaya, Mizoram, Orissa, Punjab, Rajasthan, Tamil Nadu, Uttarakhand, Uttar Pradesh and West Bengal have announced policies for setting up commercial SHP projects through private sector participation. The facilities available in the states include wheeling of power produced, banking, buy-back of power, facility for third party sale, etc. (For detailed policy please see website of the respective state).

Table 9.11

State Policies for Private Sector Small Hydropower Projects

Sl. No.	*State*	*Co-Coordinating Agency*	*Wheeling*	*Banking*
1.	Andhra Pradesh	NEDCAP	2% of energy generated	2%, 8-12 months
2.	Arunachal Pradesh	Dept. of Power	Allowed charges to be determined by SERC	With prior permission of state govt
3.	Assam	Power dept	As decided by AERC for TP. No charges for sale to ASEB	Allowed
4.	Bihar	Dept. of Energy	Allowed, terms to be decided by BSEB	Allowed
5.	Chhattisgarh	CREDA	To be decided by CSEB	Allowed
6.	Gujarat	GPCL	4% of energy generated	6 months
7.	Haryana	HAREDA	2% of energy generated	Allowed
8.	Himachal Pradesh	HIMURJA	2% of energy generated	Allowed with additional charges
9.	Jammu & Kashmir	J&K PDC	10% now, to be decided by SERC. No charges for sale to PDD local grid	Allowed for 2 months

Note: SERC: State Electricity Regulatory Commission.

NEDCAP: New & Renewable Energy Development Corporation of Andhra Pradesh Ltd.

CREDA: Chhattisgarh State Renewable Energy Development Agency.

GPCL: Gujarat Power Corporation Ltd.

HAREDA: Haryana Renewable Energy Development Agency.

HIMURJA: SHP under Himachal Pradesh Energy Development Agency.

J&K PDC: Jammu and Kashmir State Power Development Corporation.

AERC: Assam Electricity Regulatory Commission.

CSEB: Chhattisgarh State Electricity Board.

BSEB: Bihar State Electricity Board.

Source: Culled out from Indian Renewable Development Agency Compendium. *http://www.ireda.gov.in/Compendium/index%20copy*

Table 9.12

SHP Projects Installed in Private Sector (as on 31.03.2009)

Sl. No.	*State*	*Total Number*	*Total Capacity (MW)*
1.	Andhra Pradesh	41	96.93
2.	Assam	1	0.10
3.	Himachal Pradesh	33	134.45
4.	Karnataka	66	520.80
5.	Kerala	2	33.00
6.	Madhya Pradesh	1	2.20
7.	Maharashtra	4	21.00
8.	Orissa	1	12.00
9.	Punjab	10	16.65
10.	Tamil Nadu	1	0.35
11.	Uttarakhand	9	43.30
12.	West Bengal	5	6.45
		174	887.23

Source: Energy Alternatives India. *http://www.eai.in/ref/ae/hyd/hyd.html*

Watermills

Water wheels, commonly known as 'gharats', have traditionally been used in the Himalayan regions for rice hulling, milling of grain and other mechanical applications. These watermills are normally of very old design and work at very low efficiencies. It has been estimated that there are more than 1.5 lakh potential water mill sites in the country. New and improved designs of water mills have been developed for mechanical as well as electricity generation of 3-5 kW.

The Ministry is providing subsidy for development and upgradation of water mills. Local organisations such as the watermill associations, cooperative societies, registered NGOs, local bodies and state nodal agencies are being encouraged to take up these activities. A number of NGOs are now propagating watermills for electricity generation to meet small scale electrical requirements of villages. Uttarakhand has taken a lead in setting up electricity generation watermills and over 350 such watermills were installed in remote and isolated areas of the state. Nagaland has recently commenced setting up watermills/micro hydel sets for rural electrification. Watermills are also being installed in Arunachal Pradesh,

Himachal Pradesh, J&K, Karnataka and Manipur. Scheme for watermills and micro hydel projects (upto 100 kW).

Manufacturing Status

India has a wide base of manufacturers of equipment for small hydropower projects. State-of-the-art equipment is available indigenously. Thirteen manufacturers fabricate almost the entire range and type of SHP equipment. Manufacturer's capacity is estimated at about 250 MW per year. In addition, there are about five manufacturers that are producing micro hydel and watermill equipment. (See the list of SHP equipment manufacturers in the SHP section of the sitemap of the MNRE website).

Technical and Consultation Services

Consultancy services in the field of small hydro projects are available from a number of government/private consultancy organisations. The Ministry is strengthening technical institutions to provide such services. AHEC, IIT Roorkee is providing full range of technical services in the field of small hydro including survey and investigation, detailed project report (DPR) preparation, project design etc. On-site testing facility has been created at AHEC to test SHP stations for their performance. (For testing facilities please see AHEC website: *www.ahec.org.in*)

Real Time Digital Simulator for SHP

A real time simulator has been set up at AHEC which would provide hands on experience to operators of SHP stations. It is the first SHP simulator in the country. The simulator is capable of replicating all conditions of a hydropower station. AHEC is offering regular training programmes for operators and engineers of SHP stations. (For details of simulator and training programmes please see AHEC website: *www.ahec.org.in*)

Technology

Normally, small hydro projects (SHPs) are constructed in combination with water utilisation or irrigation purposes. In an SHP, the function of the hydro turbine is to convert hydraulic potential energy of water into the mechanical kinetic energy of the turbine runner, which in turn is converted into electrical energy by the generator.

Small hydro projects can be broadly classified in the following two types.

Small Hydro Projects on Hill Streams

Small streams with steep bed slopes are available in the hills, giving rise to medium as well as high head projects utilising small discharges. These schemes are normally run of the river type with a small diversion structure to divert the flows through the head regulator located in the intake portion of the diversion structure. The water conductor system would usually comprise a diversion and head regulator, a power channel, a desilting basin, forebay, penstock, power house and a tail race leading from the power house to the stream.

Small Hydro Projects on Canal Falls/Dam Toe

Irrigation canals carrying relatively high but assured discharges have several falls along their route. Small hydel projects utilising low heads can be constructed at such falls. Small hydel projects can also be located just downstream of a dam, barrage or similar structure to utilise the difference in the water level in the reservoir and in the canal downstream. A bypass channel to bypass the flows adjacent to the fall structure is constructed and the power house is constructed in the bypass channel. The bypass channel is suitably connected to the main channel.

Table 9.13

Type of Turbine for Small Hydro Projects

Type of Turbine	*Class of Head*	*Head Range for Large/Medium Sets (m)*	*Head Range for Small Sets (m)*
Pelton (impulse)	High head	above 300 m	above 150 m
Francis (reaction)	Medium head	30 to 500 m	20 to 200 m
Kaplan (axial flow)	Low head	3 to 40 m	3 to 25 m

Classification

Depending upon the capacity of the project, a small hydro project can be classified as:

Table 9.14

Classification of Small Hydro Projects

Type	*Total Capacity*
Micro	Upto 100 kW
Mini	Upto 2000 kW
Small	Upto 25000 kW

Assessment of Power Potential

Power potential can be estimated by using the following formula:

Where

P = 9.81 Q x H n

P = Power in kW

Q = Discharge in Cu. M. per second or cumecs

H = Net head in meters

N = Overall unit efficiency

The following subsidy schemes are available for the small hydropower (SHP) projects upto 25 MW station from MNRE for FY 2003-04 (Table 9.15):

The proposal is to include the following documents:

1. The subsidy will be released in one go after commissioning and commencement of commercial generation from the project.
2. The subsidy will be given through the financial institution (FI) which has provided the term loan.
3. State government administrative approval.
4. The application for the grant of subsidy, complete in all respects shall be considered in MNRE if received within six months from the date of first disbursement of the term loan or by 31.03.2004, whichever is later.

Table 9.15

Subsidy Schemes for the Small Hydro Projects

Schemes Area	*Upto 100 kW*	*From 101 kW to 999 kW*	*From 1 MW upto 25 MW*
NE Region, Sikkim, J&K, H.P., & Uttarakhand (special category states)	45% of the project cost limited to Rs 30,000/- per kW	45% of the project cost limited to Rs 30 lakh + Rs 21,625 per kW	45% of the project cost limited to Rs 2.25 crore + Rs 37.50 lakh per MW
Notified hilly regions of all other states & islands**	30% of the project cost limited to Rs 20,000/- per kW	30% of the project cost limited to Rs 20 lakh + Rs 14,400 per kW	30% of the project cost limited to Rs 1.5 crore + Rs 25 lakh per MW
Plain and other regions of all other states	20% of the project cost limited to Rs 10,000/- per kW	20% of the project cost limited to Rs 10 lakh + Rs 7,200 per kW	20% of the project cost limited to Rs 0.75 crore + Rs 12.50 lakh per MW

Note: ** The notified hilly areas would mean the hilly regions in the states, as may be notified/ dedicated by various state governments.

Source: *http://www.ecohousingindia.org/.../data/.../ECO-Housing%20final%203.pdf*

Release of Subsidy

The developer through the FI would inform MNRE about successful completion of the project, fulfillment of performance guarantee tests, testing and certificate of performance of the project by AHEC, IIT Roorkee and the commencement of commercial generation.

The developer would send monthly generation report to MNRE after start of commercial generation till the time project attains 80 per cent of the quantum of generation for the corresponding months as envisaged in the DPR, for a minimum of three continuous months. The developer would also provide proof of energy generation such as certificate from SEB regarding purchase/wheeling of power.

Concluding Remarks

Renewable energy: especially the wind energy has made good strides in India.

India is able to meet the renewable purchase obligation (RPO) of 10 per cent through the renewable energy mostly through the wind energy. The future potential of the wind, biomass and small hydropower is 4, 10 and 10 times the present level respectively.

The future potential of the solar energy in India is almost infinite and has the potential to meet the energy requirements of India. Although the bulk of the solar energy related issues: especially those pertaining to the JNNSM are discussed in the next chapter. Solar water heaters, green buildings and the urban rooftop solar photovoltaic systems are discussed in this chapter. Although the SWHs are growing fast in usage, the growth has to be manifold faster to be comparable with China.

10 International Comparison of Solar Energy and India's Journey

Introduction

Solar energy has made enormous strides in the past 5 to 10 years: especially among the top 10 countries including India. The increasing pace of the depletion of the fossil fuel-based energy resources will only make the importance of solar energy more profound.

In Chapter 9, we discussed India's endeavour to be self-sufficient in energy through appropriate renewable energy policies and implementation strategies and if possible be the world leader in production and exports of 'up-to-date' technology-based products. It is time to review India's global standing in terms of capacity, R&D efforts and take-off position concerning the solar energy using the photovoltaic (PV) and thermal technologies. This chapter also discusses the global wind power capacity and trading based on the available information. This chapter is divided into 10 sections.

Section II discusses the viability of solar energy for various domestic applications and the rapid possibility of increased viability on the grid. Section III discusses the profile of the global solar industry. Section IV discusses India's solar power industry, R&D, costs, challenges and strengths and the Central and state government's initiatives towards the solar power development. Section V to VIII discusses the Jawaharlal Nehru National Solar Mission (JNNSM), its implementation and R&D strategies and monitoring and guidance mechanism in place. Section IX discusses the global wind power profile and section X presents the concluding remarks.

Increasing Viability of Solar Energy

Solar energy is fast becoming a viable component of the power sector. The long-term benefits of solar heating and Lighting *vis-à-vis* the heating and lighting using the grid connected coal and natural gas-based power is becoming evident due to technological advances in energy conversion, cost reduction per watt of power and also due to Government subsidies. Besides, solar energy has the ability to make country energy independent and could significantly contribute to a reduction in harmful greenhouse gases (GHGs), including CO_2 and chlorofluorocarbons (CFCs), acid rains, smog and other harmful particulates.

At present *viz.*, in 2010, the world produces about 12,000 MW of solar power. It could increase a 1,000-fold by the year 2030. In India, it could increase 10,000-fold from the present level of about 9.48 MW to 100,000 MW in the corresponding period.

Financial Benefits of Solar Energy

Solar energy is a viable financial investment. (Special incentives are given to household and businesses in the form of subsidy and interest rate rebate. There is a subsidy of up to 50 per cent on a solar system in India and a tax credit on the system. In USA, 50-70 per cent of subsidy is given by the state utilities and a 30 per cent tax credit is given by the Federal government. The long-term returns on the system in the form of saved electricity tariff, rebates and tax credits are very high and even could overtake the returns on the attractive bonds and stocks. In USA, it is calculated to be even over 30 per cent.

Hedge against Increasing Energy and Electricity Costs

In India, the per unit price of electricity paid by the industry is over Rs 9 and the per unit price of the electricity paid to the power exchange by the distribution utilities when engaged in power trading is over Rs 8. By the time the power reaches the industrial customers, the price could reach above Rs 10. India cannot afford such a high electricity price which is above the price in many developed countries with the exception of California and few other states in USA.

In India, water heating contributes to over 60 per cent of electricity bill. A solar water heater costing about Rs 9,000 can nullify the monthly electricity bill to this effect and can pay back the cost of the solar heater in about two years. Any solar water heating after two years is virtually free with minimal maintenance cost. Agni Solar Systems Pvt. Ltd. and other noteworthy companies offer at least a 10-year lifetime for solar heaters.

Besides, the high electricity prices paid by the industry have a cascading effect on the prices of the commodities which has to be borne by the consumers. On top of it, higher prices also imply higher excise duty in hard currency as paid by the consumers. At the household level, solar systems can guarantee reliable power supply for at least 20 years and at the most 60 years even in hurricane like conditions.

Sinking of investment in solar system relieves the household and businesses from frequently rising power tariffs. Household/rooftop systems have the potential to generate surplus power to be sold to the grid at preferential tariff thus enabling the system become further financially viable.

Over time, surplus generation of solar power enables the securing of carbon credits rendering the system further financially viable. 20-25 years of guaranteed electricity and up to or over 50 years of actual electricity is produced by good solar power systems with minimal degradation. The system can pay for itself from the day it is installed.

Any electricity generated from beyond the date of financial solvency of the solar power system is virtually free. In other words, between the 15th and 50th or 60th year, the electricity generated is free. A recent study by ICF Consulting (funded by Department of Housing and Urban Development (HUD) and Environmental Protection Agency (EPA), in USA suggests that, energy-saving measures such as the adaptation of solar energy can add $20 to the value of the home for every $1 of yearly cost savings. This was also confirmed by the 'Appraisal' Magazine. The savings in the monthly electricity bills can be used to pay for the mortgages as well as leases thus neutralising the ownership cost. In USA, a typical 3 kW system increases the value of a home or business by up to $8,500.

Environmental Benefits of Solar Energy

'Solar is highly compatible with the values and desires of the environmentally-conscious citizen.' The study by the Department of Energy (DoE) and National Renewable Energy Laboratory (NREL) stated that 'Solar electricity is clearly a wise energy investment with great environmental benefits!' Solar energy is a sustainable form of energy. It fulfills the major objectives *viz.*, the conservation of fossil fuels, the reduction of pollutants, reduction of waste, and more efficient use of electricity. Conservation of fossil fuels is more urgent in the context of a rapidly growing economy and population and the need to secure the living of future generations.

The Reduction of Pollutants

Electricity production is dominated by fossil fuels. On the whole, the production of heat, electricity, etc., is dominated by coal, natural gas and similar fossil fuels. Coal usually constitute over 50-80 per cent of electricity production and natural gas up to 15-20 per cent respectively. In India, coal, hydro and renewable energy sources contribute to about 54, 25 and 8 per cent of electricity production respectively and the rest is produced from gas and nuclear resources respectively. The emission by coal is very high compared to petroleum and natural gas. The ratio of percentage of emission to percentage of electricity generated by source is 1.6, 1.3 and 0.94 respectively for coal, petroleum and natural gas.

There are major environmental impacts attributed to electricity generation using fossil fuels. Emissions of pollutants into the atmosphere in the form of (particulates, sulphur dioxide (SOx), nitrogen oxide (NOx), carbon dioxide (CO_2), and others) all have a grave impact on public health, water and crops. These emissions further impact many delicate ecosystems such as forests and fisheries.

Reduction of Waste

Electricity produced from coal results in a great deal of waste such as: dust from surface mining, drainage water cleaning and drying, liquid and solid waste etc. In addition to these costs to the society, electricity producers have to pay fines to the Government for perpetuating the emissions. Cleaning the emissions and waste results in greater expenditure of water, land, energy and heat. Land disposal site can be reused for solar energy or its substitute allotting alternate site for solar energy.

Over 25 years, a typical 3 kW solar energy system will offset the following in USA:

- 180,000 lbs of CO_2 (GHG),
- 600 lbs of NOx (smog),
- 500 lbs of SO_2 (acid rain),
- 300,000 miles driven in an average car (12,000 miles a year),
- It's like planting 1.5 acres of trees and
- It will prevent over 90,000 lbs of coal from being burned!

A 100,000 MW or a 100,000,000 kW which would be produced in India in the year 2030 will have great environmental benefits. Solar power would contribute greatly to keeping the GHG emissions within 550 parts per million as recommended by the Intergovernmental Panel on Climate Change (IPCC).

Note: Since the standards of emission control are lower in India and since the Indian coal has very high ash content apart from outdated technology being used in many power plants, the emission offset to be calculated can be at least 30 per cent higher than that in USA.

Positive Linkages from Solar Energy

Solar energy provides more than just clean, cheap electricity. Here is a snapshot of some solar energy benefits that are often overlooked:

Job Creation

According to the Renewable Energy Policy Project (REPP) in USA, for every megawatt of solar energy installed—35.5 jobs are created in manufacturing, installation, servicing, sales and marketing. At present, only 9.48 MW of grid connected solar power is being produced in India. If 100,000 MW of solar energy is produced in India in the year 2030, approximately 3.55 million jobs will be created. Even if we take into account only servicing, 10-15 persons are needed to maintain a MW of solar power plant. Even if that is reduced, we could safely assume that anywhere near a million jobs will be perpetuated at the level of a 100,000 MW of solar power generation.

In Andhra Pradesh, the 3,000 MW which could be produced by the year 2020 could be increased five-fold to 15,000 MW by the year 2030 and the servicing part can perpetuate over 1.5 lakh jobs. This implies that about 8 lakh people can be supported in livelihood. There is every possibility that the solar energy can exceed the target due to virtually limitless supply of sunlight and fast reducing costs and increasing heat to energy conversion efficiency with the advancement in technology. In 1971, a unit of solar energy cost about Rs 40 in absolute value. Now it costs about Rs 15-20. This costed can reduce even further. Even the land used per MW of power produced is coming down.

Conversion efficiencies of solar PV modules are demonstrated to be about 24 per cent now although practically the 18 per cent level is being implemented. Even 28 and 32 per cent efficiency is being claimed making the coal powered power plants less attractive in future. With increasing scarcity of gas and coal, the price of electricity is only going to increase making solar power more attractive in cost.

Energy Reliability

Good solar energy systems are guaranteed to last 20 years and recent studies have shown that these systems may last up to 50-60 years. Solar systems require minimal maintenance. Using the battery backup, uninterrupted power is ensured during the blackout.

Flexibility

PV systems are modular and expandable subject to the roof limit and north-south alignment of the house. One can carry them or leave them and in turn, increase the selling price of the home.

Energy Security

Energy security is manifested in the form of:

a. Reduced reliance on coal, natural gas, and other price-volatile energy sources,

b. Reduced susceptibility to energy shortages, market manipulations, and other disruptions including the geopolitical risks and terrorist threats to the sea lanes, pipelines and transportation roots and

c. Reduced stress on the existing grid.

At present in India, there is an overall power shortage of 11 per cent and a peak time shortage of 15 per cent which is only expected to increase further. Rooftop solar systems connected to grid can reduce the distance between the production and consumption points thus minimising the transmission and distribution losses.

Developmental Benefits: Sustainable Development due to Solar Energy

Sustainable development meets the needs of the present without compromising the ability of future generations to meet their own needs. Energy has a very close relationship with all the three dimensions of sustainable development:

1. The environment,
2. The economy and
3. Social welfare.

Solar energy positively affects each of these dimensions of sustainability. Solar energy offers unique and valuable benefits for a sustainable energy future.

1) Clean energy,
2) No unwanted emissions or noise at point of use,
3) Very little environmental impact,
4) Modular and scaleable *viz.*, can upgrade or move at any time and
5) Available in virtually any size.

Solar energy has a higher chance of replacing fossil fuels *vis-à-vis* the other forms of renewable energy. Apart from the fact that the sun's power is limitless in the foreseeable future *viz.*, at least for another two billion years, solar energy systems are easily installable on rooftop and on an industrial scale. The system can be easily integrated with power grid from the rooftop as well.

Non-Energy Benefits of Solar Energy

There are some substitutional benefits associated with solar energy.

1. Electrical

a. Contributes to the growth in the power sector.

b. Reduces demand for utility-based electricity.

c. Ensures power in times of emergency.

d. Enables grid support for rural lines.

e. Increases electricity efficiency by reducing transmission and distribution losses that usually occur in the long power lines.

f. Improves grid reliability and resilience across long distances.

Note: If rooftop systems are installed, the point to point transmission distance gets reduced and transmission and distribution losses can thus be reduced. If the smart grid infrastructure with the transmission capacity of over 1,200 kV is implemented, there could virtually be no transmission and distribution loss. Besides the smart grid infrastructure is more durable and stable.

2. Architectural Substitutes certain Building Components

a. Provides waterproofing, fire protection, wind protection, acoustic control, day lighting, shading, thermal collection and dissipation.

b. Aesthetic appeal through color, transparency, non-reflective surfaces.

c. Reduced cost of building maintenance and roof replacements.

Note: If architectural benefits are included in solar energy benefits, new houses could become more attractive with the installation of solar system. Industry and government should project this aspect while promoting consumer awareness in favour of solar energy. Besides, the architectural benefit imbibes insulation contributing to anywhere between 25-40 per cent of reduction in energy use. In Indian context, insulation is more important than fireproofing because Indian constructions are concrete based unlike the wood-based constructions in USA and other developed countries in northern hemisphere.

3. Socioeconomic Benefits

a. Creates new industries (especially local), products and markets.

b. Creates local employment in installation and servicing.

c. Reduces fuel imports and price volatility.

d. Supports urban renewal and rural development.

e. Decreases adverse environmental impact, social dislocation and infrastructure requirements.

Note: With the availability of reliable power in distant areas, the developmental benefits of education, industries, employment and quality of life can become more universal. The infrastructure requirement in the form of expensive power lines in low population density areas can be minimised. Reduced adverse environmental impacts due to the use of fossil fuels can save money for the Governments in the form of reduced health expenditure. This money can be used to provide subsidy, tax rebate and interest rate rebate on solar power systems.

f. Reduces the cost of fuel transport and pollution from the use of fossil fuels in rural areas.

Note: The density of power lines is very low in rural areas because the population density could even be 1/50th of that in urban areas. Besides, the use of diesel pump sets and excess fertilisers due to inadequate and untimely availability of water for irrigation are harming the productivity of land and livelihoods of the people.

g. Reduces the risk of nuclear accidents.

Note: The Chernobyl Nuclear accident in 1986 severely affected the biodiversity in about 60,000 sq km of the land area or in 1/2,200th of the earth's land area. Had the accident been larger, the damage would have been colossal affecting manifold population and land area. Even in a simplistic sense, 60,000 sq km or 15 million acres of land area can produce about three million MW of solar power or about 18 times the present production in India. The self-sufficient production of eight lakh MW can thus be multiplied four-fold using the present technology of 18 per cent energy conversion.

The state of Rajasthan has set aside about nine million acres for establishing solar power plants which is more than sufficient to meet India's requirements of 2030.

h. Creates a symbol for sustainable development.

i. Presents potential for international cooperation, collaboration and long-term aid to developing countries.

Promoting the Use of Solar Energy

At the outset, solar power should be made cost-effective *vis-à-vis* the conventional energy. The entire lifecycle of a system needs to be understood before suggesting a financial plan to the customer. Solar system producers in US and other developed countries are pursuing a system of solar leasing or a system of power purchase agreement (PPA) with the customers.

Solar Leasing

Solar leasing is based on the 20-25 year guarantee period for the system. Under this plan, customer is simply the tenant of the system. Once a customer chooses to disown the system, she/he can simply walk out.

Power Purchase Agreement

Power purchase agreement (PPA) is the system by which a solar system is purchased on finance. The installment repayment is especially beneficial for the customer because there is high amount of subsidy that can be availed on installments and even the tax credits can be availed on installments over the guaranteed lifetime of the system *viz.*, 20-25 years. In contrast, a 'one time' subsidy of 50-70 per cent and tax rebates for a shorter period of 12-15 years is burdensome to the governments. Through the PPA at the customer level, governments can spread the subsidy and tax rebates over a period of 20-25 years. Besides, governments or the allied agencies like the Indian Renewable Energy Agency (IREDA) can find ways to minimise the burden of interest rate subsidy.

Spread of Solar Energy in India

In India, the preferential tariff of anywhere between Rs 15 and 18 is being envisaged along with the PPA of 25 years. Even the interest rate subsidy is given to the tune of 7 per cent by IREDA making the effective 5 per cent. The introduction of renewable energy certificates will make the solar power producers earn carbon credits to the tune of about $17 per tonne of emission prevention as per the present price. These benefits should be transmitted to the customer through PPA and solar leasing.

In case of grid connected large scale solar power plants, customers should be encouraged to purchase shares of the solar power plants so that they can reap benefits throughout the lifecycle of the plant and the infrastructure. To purchase the shares, IREDA and other financial institutions should give loans to people at preferential interest rates with due support from the Government. Only then, solar power can match and even overtake the power produced using fossil fuels in absolute MWs.

Solar Care Warranty Programme

Solar system producers are providing the lifetime warranty for labour maintenance and servicing. Packages like the 'geo solar' are meant to spread the cost of maintenance and servicing over the lifetime of the system.

Summing-up

On the whole, the solar power bestows substantial financial, environmental, economic, architectural and social benefits to people. The extent of benefits is largely unaccounted for by the studies and a systematic and contextual effort has to be made to capture these benefits in quantitative values. People and governments should be made aware of these benefits and costs so that the production and use of solar power can be spread using appropriate policies and instruments.

Profile of Global Solar PV Industry

Production

The production of solar PV modules in the year 2008 was estimated at about 7.3 GW. The total production increased by about 84 per cent above the 2007 level.

During the year, Chinese and Taiwanese production increased more than proportionately with China ranked number one with about 2.4 GW. Europe ranked next with 1.9 GW, Japan next with 1.2 GW and Taiwan next with 0.8 GW respectively.

Figure 10.1

Trends in Solar PV Production (MW)

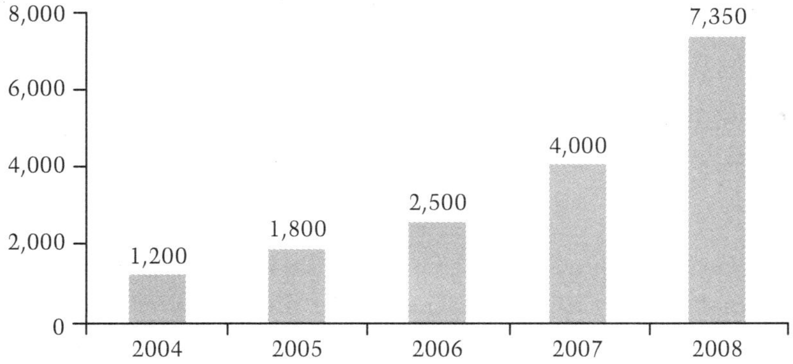

Source: Navigant.

Installations of Solar PV across the World

The market for installed PV systems doubled in 2008 from 2007 and was estimated at 5.7 to 6 GW. It further doubled to 12 GW by the beginning of 2010. The impressive growth in 2008 is attributed mainly to an exceptional growth in the Spanish market, which rose nearly five-fold from 560 MW to 2.7 GW. The next largest and most stable market was Germany with 1.9 GW followed by the US with 342 MW, South Korea with 282 MW and Italy with 258 MW and Japan with 230 MW respectively.

Market Profile for Solar PV Industry

Europe

The European PV production grew by over 80 per cent and reached 1.9 GW. Yet Europe continued to be the net importer of solar cells and modules. The rapidly developing markets in Europe apart from Germany and Spain include the Czech Republic (51 MW), France (46 MW) and Portugal (50 MW) respectively. The growth is becoming more rapid in other regions which could possibly reduce the market share of Europe.

USA

USA was the third largest market with 342 MW of which 292 MW was grid-connected. California, New Jersey and Colorado accounted for more than 75 per cent of the US grid-connected PV market. The US Senate voted to extend the tax credits for solar and other renewable energies in 2008.

South Korea

South Korea became the fourth largest PV market in 2008 by increasing the capacity over three-fold between 2006 and 2007 from 25 MW to 78 MW with plans to add a further 500 MW by 2009. With a five per cent share of new and renewable energy sources by 2011, solar PV capacity would be 1.3 GW by 2012 and four GW by 2020 respectively. The 2009 National Renewable Energy Plan focusses on new investments in the infrastructure and technological developments concerning renewable energy and progressively increasing the share of renewable energy by 4.3, 6.1 and 11 per cent by 2015, 2020 and 2030 respectively.

Towards this endeavour, South Korea introduced an attractive feed-in Tariff for 15 years, along with investment grants up to 60 per cent. From 2012, renewable portfolio standards would replace the feed-in tariffs. The new tariff scheme enables the choice of a 15-year guarantee and a higher kWh price or a 20-year guarantee and a somewhat lower kWh price.

The previous 100 MW cap was enhanced to 500 MW which is to be achieved in 2009. 1,00,000 houses and 70,000 public/commercial buildings are to be equipped with PV systems by 2012. Some of the larger projects will qualify for clean development mechanism (CDM) credits enabling the trading in Certified Emission Reductions including carbon reductions.

China

The Chinese PV market more than doubled in 2008 to 45 MW. The However, the domestic market was less than 2 per cent of total PV market. A drastic change is expected with an RMB 210 billion (US $22 billion or Rs 1,00,000) allocated for green energy programmes as part of the RMB 4 trillion stimulus package announced in March 2009. A large scale programme for the development of domestic grid-connected solar power was announced in 2009 with an RMB 20 billion or Rs 100 billion. At the end of 2008, China's cumulative installed capacity of Solar PV power was 150 MW of which 55 per cent was from standalone PV power generation units. Solar water heater collector area was 125 million square metres accounting for 60 per cent of the world's market. China plans for a 12 GW solar capacity in 2011. Under the new energy stimulus plan announced in July 2009, China's 2020 targets for installed solar capacity is 20 GW.

China's 'Golden Sun' programme aims to add 642 MW capacities by 2012 through the approved 294 projects. Ninety per cent of the installed capacity of 2,400 MWp is exported by China. Chinese government will co-finance the utility scale projects by 50 per cent and remote area projects by 70 per cent at the generation, transmission and the distribution levels. The surplus power from the standalone units should be purchased by the utilities at the benchmark of the tariff paid to the coal-fired power plants on a kW basis.

Taiwan

Taiwan aims at subsidising the manufacturers and consumers. The focus is on the Thin Film production with about eight producers setting up their own plants.

Japan

Japan's growth was comparatively slight at 9 per cent in 2008 to 230 MW compared to a two-year decline. Japan's Ministry of Economy, Trade and Industry announced the capital subsidy for household and small-scale users and a feed-in-tariff for commercial investors. An additional 400 MW would be generated through 1,00,000 systems.

Future Global Trends

From the present level of 12 GW, the global installed capacity is expected to reach 54 GW by 2012 at an expected annual growth rate of over 50 per cent. While the global capacity would increase 4.5 times, much of this increase depends on China's realisation of its ambitious target of almost 17.5 GW. China's industry should grow 7.5 times with an annual growth rate of about 80 per cent.

Trends in Global Solar PV Prices

The prices witnessed downward trend till 2005. Between 2005-2008, the supply of polysilicon fell short of the demand causing the price rise. During 2008, the prices were US \$3.5 per watts peak (Wp). Inventory increased to two GW in 2009 due to the economic slowdown in 2008. Towards the end of 2009, the prices fell to $1.9 to large and $2.5 to medium level buyers. Since the inventory fell to 500 MW and the market grows in the traditional and also in the new markets, the prices will only recover further in spite of certain countries providing subsidy for the manufacturers and for the end users. In the context of India, certain key components like the 'evacuated glass tubes' have to be imported. The solar heaters of similar capacity costing about $400 in India cost only below $200 in China.

Solar Power in India

The capacity of solar PV-based power in India is estimated to be about 750 MW for cells and about 1,250 MW for modules at the end of 2010. However, the cumulative production was 1,753 MW of cells and 2,403 MW of modules at the end of 2008. A large proportion of the production was exported. By the end of the third phase of the Jawaharlal Nehru National Solar Mission (JNNSM), the annual domestic production capacity is to be increased to 8/10 GWp.

The field is dominated by 15 companies led by Moser Baer and Tata BP solar. Several projects in the pipeline could not take off and several are in the offing. Thar Desert would become the major manufacturing hubb for Solar PV modules. In 1996, Amoco/Enron Solar Power Development planned to build a 50 MW solar PV plant in the Thar Desert near Jaisalmer in Rajasthan. In addition, a 50 MW PV plant and a 200 MW solar chimney were proposed. Neither the Amoco/Enron plant nor the proposed plants were completed.

The Government of Rajasthan has set aside 35,000 sq km area of the Thar desert for solar power development. Astonfield Solar (Rajasthan) Pvt. Ltd., promoted by Astonfield Group of Companies is developing the first solar PV project in Rajasthan. It was commissioned in early 2010.

Relevance of Solar Energy for India

India receives abundant solar radiation thanks to the proximity to the equator, with the peninsula enjoying over 300 sunny days annually. The cost of solar power declines progressively making it more attractive form of renewable energy. Various factors leading to a decline in the cost of solar energy include: the setting up of large-scale plants, integration across the value chain, declining cost of raw material, reduction in the material consumption and higher efficiency of modules. These advantages enable the solar power to progressively eliminate the subsidies on kerosene—a domestic oil that causes domestic pollution although not to the tune of the firewood.

Unlike many sources of electricity, solar power has many direct applications *viz.*, in heating and cooking systems, in lighting, in powering electronic appliances and gadgets etc. Direct application is well focussed in India's government policy and is being actively supported budgetarily. Recently, grid-connected solar power applications are being given special impetus and appropriate technologies including the solar PV are being developed and implemented.

Overview of Solar PV Industry in India

Solar Cells and Solar PV Production

During 2008-09, 15 companies were manufacturing solar cells, 20 companies were manufacturing PV modules and 50 companies were manufacturing various kinds of PV systems respectively. By 2010, about 700 MW of solar cells and about 1,750 MW of solar PV modules are manufactured in India.

Foreign Trade of Solar PV

India is a net exporter of solar PV modules. Sixty-six per cent of the total production was exported in 2010. During 2008-09, about 80 per cent of India's solar PV imports were from Germany and Taiwan and the rest were imported from Spain, China and Japan. Germany and Spain are the major export markets for India with 75 per cent of the total exports. The rest of the exports are primarily made to the Netherlands, Belgium, Italy, Norway, France and Afghanistan respectively.

Solar PV Applications in India

Globally, grid connectivity accounts for nearly 75 per cent of the installed capacity and off-grid lighting and consumer applications account for 25 per cent respectively. PV installations in India mostly comprise off-grid systems and small capacity applications mostly used for public lighting including streetlighting, traffic lighting and domestic power backup in urban areas and small electrification systems and solar lanterns in the rural areas. Water pumping is becoming a thrust area of application. Government agencies like the Indian Railways are becoming the major consumers.

Research and Development (R&D)

Solar technologies used in India are not up-to-date. In fact, the technologies are still in their developmental stage. Ninety per cent of the solar modules are manufactured using crystalline silicon (cSi) technology and 10 per cent are manufactured using thin film or amorphous silicon (aSi). Thin film technologies produce more power per unit area than the crystalline technology.

Future Trends of Solar Power in India

India has got the twin advantage of high population density and clear sunshine. The large population in the long run can sustain the momentum of handsome economic growth. Clear sunshine and the evolving solar power technologies can sustain this growth without adding to or by substantially reducing carbon emission. India needs to provide energy security on a sustainable and cost-effective basis.

India's NAPCC and the JNNSM advocate the installation of 20,000 MW solar power generation capacities by 2020 to alleviate the power shortage in India. The present power shortage is estimated at 11 per cent off-peak and 15 per cent peak. The achievement of the targetted generation of solar power and the targetted generation of renewable energy-based power as a whole would greatly reduce the power shortage in India.

The main technology to generate solar power is the PV technology. The key challenges to solar PV installation are: achieving grid parity, high cost and scalability.

a. *Grid parity*: Although India is endowed with 300 days of clear sunshine, it varies across areas as per the difference in the climate and the length of the monsoon season. The collection capability of the solar panels varies highly across space and time. This high degree of variability causes the problem of reliability due to voltage drop and power supply disruption at the tail end of the grid as well as in the non-grid connected establishment. At least 33 kV production capacity should be maintained by a solar power plant and 33 kV should be transferred to the grid to achieve grid parity. Achievement of grid parity is likely to be a reality by the year 2022.

b. *Cost*: Solar energy is the costliest in India, and in the world. As stated, a unit of solar power cost up to Rs 30 in India. This will come down to the level of domestic coal-based power if the global production of solar power increases 16-fold and if appropriate technologies are developed and deployed by the year 2030.

c. *Scalability*: Five trillion kW of solar power can be produced in India if 300 days of clear sunshine is available. However, availability of land is the key challenge to achieving this target. Approximately 1 sq km of land is required to generate 20/60 MW of solar power. Keeping this in mind, to generate 800 GW of power in India by 2050, 14,000 to 40,000 sq km of land has to be allotted for solar power generation. This is next to impossible with land being highly scarce in India which needs to be put to highly competing uses like agriculture, industrial development, infrastructure projects etc.

Under these circumstances, the immediate next best alternative to solar PV-based power plant is the highly distributed grid-connected rooftop solar PV installations. The latter option is extremely expensive and does not achieve economies of scale. Whether grid-connected or not, if the solar PV modules become smaller and cheaper overtime and if the technology is effectively able to compete with conventional fossil fuel-based technologies, the rooftop installations can proliferate and will enable the realisation of JNNSM targets easily.

But, here is a catch. The real cost of Solar power would come down slowly while the real cost of fossil fuel-based power could increase faster. Already a unit of coal-based power is priced at Rs 7 at the base load and Rs 8.5 at the peak level. A unit of diesel-based power cost about Rs 15 which is much nearer to the cost of solar power. The reliance on diesel and imported coal can increase the real cost much faster than anticipated in India.

We should understand that the 'Integrated Power Policy' of India allows 20 per cent of coal to be traded in the 'spot' market and the Government of India is envisaging the total decontrol of the prices of fossil fuels *viz.*, the petroleum and gas. Further, the cost of environmental degradation is getting incorporated in the price of coal.

Most of the coal mines in India are operated by the public sector. Hence, labourers tend to resist greatly to technological changes. Hence, attempts to reduce the real cost of power using more advanced technology for mining and processing would be thwarted by frequent labour agitations.

The very existence of coal mines in the Maoist/Naxal dominated areas gives impetus to this resistance. Under these circumstances, research and development should be given greater impetus to bring down the real cost of solar power much faster than the increase in the real cost of fossil fuel-based power.

Cost of Power from Solar PV in India

The Central Electricity Regulatory Commission (CERC) assumes the capital cost of the plant at Rs 17 crore per MW (including the cost of BoP, etc). The useful life of the plant is assumed to be 25 years. At 19 per cent capacity utilisation, the per MW generation of electricity would be about 1.66 million units annually. The internal rate of return for the first 10 years is calculated at 10 per cent. The per kW cost of electricity could be Rs 23.78, 14.74 and 11 in the 1st, 10th and in the 25th year of the plant life assuming a debt/equity ratio of 70/30.

Challenges for India's Solar PV Industry

The lack of domestic base for raw material, technology, skill set and poor demand for solar PV-based products are the key challenges for the growth and development of the industry in India. India has to import the basic raw material *viz.*, the silicon wafers.

The spot prices are less than ever. However, the past contracts assuming the price rise are hindering the expansion plans of several firms. As a net exporter, Indian firms lost orders due to international economic slowdown between 2007 and 2010. The domestic market is not yet mature to use larger high capacity application and hence, the domestic firms would need more time to take-off. As the world leader, China's solar cell production is six times that of India and Taiwan's is double that of India making them more cost competitive on account of the economies of scale enjoyed by their firms *vis-à-vis* those of India.

The rate of commercialisation in India is low due to low consumer awareness and less focus of the Governments on tax reduction, capital subsidy and rebates prior to 2009.

The taxes on domestic products is more due to duties of 12.8 per cent and 21.5 per cent on the imported cells and wafers *vis-à-vis* zero duty on the import of the completely built modules. Besides, strengthening of the knowledge and the value chain and the provision of the quality infrastructure *viz.*, the continuous supply of water and power in the manufacturing zones will enhance the quality, quantity and profitability of the manufacturers along with the availability of quality and affordable products to the consumers.

Strengths of India's Solar PV Industry

The conversion services offered by India are cost-effective making even the advanced countries avail India's services. India can offer lot of manpower for 24 hour plant operation. A crystalline silicon manufacturing firm requires 90 technical persons to operate a 50 MWp/year capacity wafers-to-modules manufacturing plant. But, only 45 technical persons are sufficient to manufacture thin film-based modules in a plant with a capacity of 50 MWp a year.

Government Initiatives to Promote Solar Manufacturing in India

Government has taken several initiatives to promote the penetration of solar power in India.

Generation-based Incentive through the National Rural Electrification Policy, 2006

The policy aims at providing electricity to all the households in India. A minimum 'lifeline' level of consumption of 1 unit (kWh) per household per day is to be assured by 2012. Off-grid solar PV applications are part of this policy to supply power in the regions where the grid cannot reach. The generation-based incentives (GBI) announced in January 2008 allows a state to develop up to 10 MWp capacity and a single developer up to 5 MWp with 1 MWp capacity at a single location.

Up to Rs 12 per kW can be provided as GBI through the IREDA depending on the tariff announced by various state governments. The key government agencies, their roles and responsibilities are discussed in the previous chapter.

Semiconductor Policy of 2007

The semiconductor policy is meant to encourage semiconductor and ecosystem manufacturing. It offers a capital subsidy of 20 per cent for manufacturing plants in special economic zones (SEZs) and 25 per cent for manufacturing plants outside of SEZs. To qualify for the subsidy, the net present value (NPV) of investment should be US $212 or Rs 10,000 million.

Demonstration Programme on Tail-End Grid-Connected Solar Power Generation

The tail-ends of the grid in rural areas experience voltage fluctuations and power outages. A solar PV plant connected at the tail-end of the grid improves the quality of power. Under this scheme, generation companies and state nodal agencies are entitled to set up grid-connected solar PV plants between 25 kW and 1,000 kWp capacity. The scheme provides 50 per cent of the basic cost of the plant, subject to a maximum cap of Rs 10 crore per MWp. The support will be available until the setting up of four MW capacity in the Eleventh Plan.

Rural Electrification

MNRE is providing solar PV-based lighting and heaters and solar panels for household use through the Indian SPV programme for decentralised applications of photovoltaic technology under the Rajiv Gandhi Grameen Vidyutikaran Yojana (RGGVY). Far-flung villages not connected by the grid would benefit from the scheme. MNRE conducts the demonstration and use of solar PV systems by supporting streetlighting, household lighting, water pumps, solar lanterns, heaters, school computers stand-alone solar PV plants and other systems mainly in rural areas. Up to 90 per cent subsidy is given for solar power systems subject to benchmark. The solar streetlights grew by about 109 per cent between March 2008 and March 2009 with the installed number of lights crossing 1,19,000 in India.

Rural Solar Thermal Demonstration Programme

MNRE supports the demonstration and use of solar thermal technologies *viz.*, cookers, dryers, stills, heaters etc. Government supports entrepreneurs and consumers subsidising the testing, certification and the purchase cost. Up to Rs 1,500 for dish type cookers and up to Rs 15,000 for community cookers are subsidised by the Government. Up to 50 per cent subsidy is provided for solar dryers and heaters in rural areas. Solar dryers are used to dry mango pulp, bars, fruits and fish.

The marketing of solar thermal-based products are being carried out by Akshay Urja shops and self-help groups. Also repair and after sales services are being provided by Akshay Urja shops. The growth rate of solar cookers has markedly slowed down with the growth being only 2.5 per cent or 15,000 between March 2008 and March 2009. However, increasing impetus is given for community cookers by the temples *viz.*, the Shirdi Sai Baba Sansthan and the Tirumala Tirupati Devasthanam along with other temples and community organisations feeding many thousands of people every day.

Apart from the above three programmes, MNRE supports wind-solar hybrid systems in the range of 1 to 10 kW in areas where the annual average wind speed is 15 km per hour at the height of 20 metres. The projects are being implemented actively by the concerned state nodel agencies in Meghalaya and Karnataka. In Meghalaya, the Meghalaya State Non-Conventional and Rural Energy Development Agency is using a 5 kW system to power its office space. In Karnataka, streetlighting is being provided using the systems.

Solar Energy in Urban Areas

Solar energy programme for urban areas encompasses solar water heating systems for domestic, industrial and commercial establishments; solar air heating/steam generating systems for community cooking and industrial applications; solar buildings incorporating active systems and passive designs, the national rating system for green buildings (GRIHA) to promote solar/green buildings; Akshay Urja shops for sale and servicing of solar systems; solar photovoltaic products and devices for streetlighting, hoardings, traffic signals, power packs etc.

Government provides publicity support to manufacturers of solar water heating systems and interest free loans to the consumers in the northeastern states, Jammu and Kashmir, Himachal Pradesh, Uttarakhand, Jharkhand, Chhattisgarh and Andaman & Nicobar Islands. To other consumers, commercial and industrial organisations, a soft loan with 2, 3 and 5 per cent interest rate along with an incentive of Rs 100 per square metre of solar heater installation is provided. If soft loans are not availed, capital subsidy is provided at the rate of interest subsidy.

Eighteen states and 41 municipal corporations in seven states have resolved to provide support to green buildings. Six states propose to provide a property tax rebate of 6 to 10 per cent, if solar heaters are installed. A monthly electricity bill rebate of up to Rs 150 can be provided. Bangalore leads the way in India in rooftop solar installations—primarily water heaters. The total installed capacity is 200 MW. Bangalore along with Pune makes rooftop solar installation mandatory for all the new structures. Bangalore goes further by awarding a monthly rebate of Rs 50 on electricity bills if the rooftop solar installations are set up. Bangalore is planning to transfer the surplus rooftop solar power to the grid by paying the households.

Green buildings can save up to 30-40 per cent of conventional fuels used in home lighting and heating. A national rating system (GRIHA) is developed to justify the setting up of green buildings and to prepare passive architectural designs suitable for homes and commercial establishments in various climatic zones. MNRE provides 50 and 35 per cent subsidy for institutions and industries installing solar air heating and steam systems respectively. While air heating systems enable the drying of fruits and pulp, steam systems enable large-scale community cooking and heating. 284 Akshay Urja/Aditya solar shops are functioning in India to sell solar-based products and devices and to provide repair and after sale services.

Solar/green cities programme was launched to encourage and assist urban local bodies to assess energy consumption status, to set clear targets for reduction of projected energy demands by up to 10 per cent, to prepare action plans for energy generation using renewable sources of energy, and conservation of energy used by Government establishments. MNRE provides funds for conducting assessments and for the preparation of Master plan.

During the Eleventh Plan, 60 cities are proposed to be converted as 'solar cities.' Of which, 15 cities *viz.*, Agra, Moradabad, Rajkot, Gandhinagar, Nagpur, Thane, Kalyan-Dombivli, Indore, Imphal, Kohima, Dehradun, Chandigarh, Gurgaon, Vijayawada and Coimbatore are approved to be developed as solar cities.

State Level Initiatives

Solar PV systems are installed under the aegis of state nodel agencies, municipal corporations and traffic police departments. Haryana, Maharashtra and Uttar Pradesh lead the way in the installation of solar power systems and devices closely followed by West Bengal. Most of the installations are for streetlighting. Mandatory installation of renewable energy technologies and systems are being pursued by Government. Building being constructed for the usage in Commonwealth Games in Delhi; premier institutions viz., the IIMs etc., the Damodar Valley Corporation office in Kolkata and several other industries and power plants are setting up the systems. Mandatorily solar heaters in Government guest houses is being adopted.

Some of the most important state level initiatives are presented here.

Andhra Pradesh

The Government of Andhra Pradesh is developing a solar farm cluster called solar city on a 10,000 acre land at Kadiri in Anantapur district. Solar city would attract Rs 3,000 crore investment in the first phase. Four firms (Sunborne Inc., Lanco Solar, AES Solar and Titan Energy) have signed MoU with the state to set up plants with the combined capacity of 2,000 MW.

Karnataka

Karnataka Power Corporation Ltd. (KPCL), has implemented two tail-end projects—each with 3 MWp capacity and has awarded a third project with 3 MWp capacity in 2009. The plants are established under the Arunodaya scheme for ensuring assured power supply to rural areas to power the irrigation pump sets.

Gujarat

Gujarat's policy aims at harnessing the solar power from the wastelands. Besides, the initiatives to promote R&D and technology transfers along with the promotion of environmental awareness, skill development and employment generation for local youth become significant. The incentives for 25 years can be availed before March 2014. The minimum plant capacity should be 5 MWp. A two per cent wheeling charge is imposed to safeguard the interest of the state utility. Of the 716 MWp of approved projects, 365 MWp of solar PV would be through 24 projects and 351 MWp would be through 10 solar thermal projects.

Rajasthan

The state government has announced two major schemes to promote solar energy applications.

MNRE's Remote Village Electrification Scheme

Under this, domestic lighting systems (DLSs) are distributed to remote villages. MNRE will subsidise 90 per cent of the benchmark cost of Rs 12,500 per DLS that excludes Rs 1,250 for operation and maintenance for a period of five years. On the whole, the state government and consumer have to pay Rs 3,335 and Rs 1,500 respectively implying the total cost of Rs 16,080.

Rajasthan Rural Electrification Programme

This programme aims at the distribution of streetlighting systems and the domestic lighting systems in the remote areas and ensures financial assistance from the state and Central governments and the MNRE. The programme subsidises the SPV water pumps under the MNRE programme by Rs 30 per watt of SPV module capacity, subject to a maximum of Rs 50,000 per pumping system.

West Bengal

For demonstration programme on the grid-connected solar power generation, West Bengal Green Energy Development Corporation Limited (WBGEDCL), Kolkata is setting up a 2 MWp capacity solar PV power plant at Asansol, West Bengal. The project was commissioned during 2009-10.

The Jawaharlal Nehru National Solar Mission (JNNSM)

The JNNSM aims at establishing a solar power capacity of 20,000 MW by the year 2022. For this purpose, the period is divided into three phases and appropriate targets are set. To realise this target, appropriate policy formulation, implementation and monitoring mechanism, evaluation procedures and feedback mechanism is being envisaged.

Target Setting

During the three phases, solar power capacity is expected to be augmented by at least 1,000, at least 3,000 and up to or over 16,000 MW.

Policy Formulation Strategy

The capacity addition should be carried subject to mid-term and end-term review of all the phases. The phases are divided as 2009-2013, 2013-2017 and 2017-2022 respectively.

In the first phase, the policy should aim at the proliferation of the use of electricity throughout India including in remote far-flung areas using 'low hanging fruits' *viz.*, the off-grid solar power devices and systems. Simultaneously, the policy regime should support R&D and design of the grid-connected solar power systems.

In the second phase, grid-connected solar power systems would be installed after a detailed review of capacity trends and the cost of domestic and internationally available systems and technologies. Through periodic cost and technology reviews, Government should be protected from subsidy exposures and IREDA that refinances the banks and non-banking financial companies (NBFCs) involved in the financing of the purchase and manufacturing of solar power devices and systems would be able to finance R&D projects instead of supporting the manufacturers using the existing technologies.

There is a possibility that over 10,000 MW of grid-connected solar power would be available in the second phase and the chance of reaching or crossing the 20,000 MW limit in the third phase depends greatly upon the technology, cost and system scalability factors at the end of the second phase.

If things go well, grid connectivity would be easily achieved by the year 2022 and cost parity with domestic coal-based power would be achieved by the year 2030.

Implementation Strategy to be Envisaged in the Solar Energy Policy

Organisation

An autonomous Solar Energy Authority embedded within the mission under MNRE should enable technology development, provision of incentives, manage funding requirements and setting up and demonstration of pilot projects. The authority is directly answerable to the 'Prime Minister's Council on Climate Change' on the status of the programme.

Projections

During the first phase, existing proven technologies and systems employing low temperature-low performance solar collectors below 80 degree centigrade should be promoted for usage. During the second and third phases, a high degree of energy efficiency should be achieved. At the end of the first, second and third phases, the off-grid solar power capacity should be increased to 200, 1,000 and 2,000 MW respectively. By the end of the first, second and third phases, 7 million, 15 million and 20 million sq m of solar collector capacity should be installed in the country. This implies that the energy conversion efficiency of the solar collectors should be increased over three-fold during the three phases.

While 15 million sq m of solar collector area is expected to produce 1,000 MW of power by the end of the second phase, the additional 5 million sq m of area is expected to produce further 1,000 MW taking the cumulative total to 2,000 MW of power. This has the ramification in the land usage. Only 1/3rd of the land used today at the end of the third phase could produce the amount of output produced at the end of the second phase.

Taking the first phase into account, the power production can be increased 10-fold by augmenting the land use only less than three-fold. It is a major R&D challenge for India's government and the private corporate sector. However, the increase of solar collector area would not be difficult taking into account the current rate of growth. India has a total solar collector capacity of 3.5 million sq m by March 2009. This is about 30 per cent more than the capacity of 22 million sq m in March 2009. Assuming even a 20 per cent growth rate, the target of 15 million sq m is easily achievable by the end of the second phase and the target would be much exceeded by the end of the third phase taking the total collector area to above 42.5 million sq m.

Implementation Mechanism

To realise and augment efficiency in solar technology, a synergy between the proliferation of the usage and augmentation of manufacturing capacity should be achieved. The existing programmes and schemes under the Rural Village Electrification (RVE) and others supported by MNRE should be continued and bolstered. The active support for solar power installations *viz.*, the 90 per cent subsidy for the rural and 50 and 35 per cent subsidy for the urban domestic and commercial users and subsidy amounting up to the level of interest rebate provided in various categories of states and union territories is expected to vastly augment the manufacture and use of solar power systems and devices in India. Some of the successful solar power uses include: the use of management information system (MIS) in Madhya Pradesh, *tussar* reeling by women self-help groups in Jharkhand, milk chilling plant in Gujarat and running of cold chains. Special incentive packages should be provided to the company scale and rooftop manufacturers.

Power utilities such as the state electricity boards should purchase power from the solar power plants under the Renewable Purchase Obligation (RPO) as envisaged in the Electricity Act of 2003. From 0.25 per cent, the RPO of solar power should be increased to 3 per cent by the year 2022. The National Thermal Power Corporation (NTPC)'s Vidyut Vyapar Nigam Ltd. is authorised to purchase power from the plants able to transfer at least 33 kV of power to the grid.

Rooftop power producers should be able to transfer at least 11 kV to the grid on a sustainable basis. The price paid to the producers will be in accordance with the RPO of the respective state governments. If the electricity boards do not have ready money, refinance facility will be provided by IREDA. Manufacturing capacity of the solar PV modules in the existing 15 dominant companies should be bolstered from the present level of 700 MW to 8/10 GW to suffice for domestic use and to enable exports.

Especially, the solar silican material and films should be manufactured on a larger scale and appropriate R&D should be undertaken and if necessarily, technology transfer should be facilitated through refinance by IREDA. Government should reduce its subsidy exposure in concurrence with the reduction in the cost of solar power devices and systems. By the end of the second phase, the maximum subsidy level should be capped at 30 per cent.

R&D Strategies

The R&D strategy would focus on five categories.

i) Basic research with long-term perspective towards the development of new materials and appropriate innovations on new processes and applications.

ii) Applied research aimed at improving the existing processes, materials and technology to enhance performance, improve durability and cost competitiveness of the systems/devices.

iii) Technology validation and demonstration projects aimed at field evaluation of different configurations including hybrids with conventional power systems to obtain feedback on the performance, operability and costs.

iv) Development of R&D infrastructure with public-private partnership (PPP).

v) Support for incubation and startups.

The Focus

The focus of the three phases should be to develop medium and high temperature applications maximising the energy conversion efficiency of solar collectors. IREDA should support the R&D through refinance and technology transfer from foreign sources should be inbuilt in the R&D strategies. A programme known as 'R&D for Solar India'—creating conditions for research and application—should be initiated to focus on: improving the efficiency of the existing materials, devices and applications and to reduce the costs of the systems. Further, the programme envisages to develop new applications integrate and optimise the power supply.

Cost-effective and storage augmenting technologies should be developed through the programme. This in turn enables the overcoming of the space constraint and mitigates the variability problem using better concentrators, applying of nanotechnology and the use of better and improved materials. Best technologies will be able to survive under the market conditions in the programme due to its technological neutrality.

The Scheme

A solar Research Council should be set up to review the existing research projects, future research possibilities and international cooperation and collaboration. A comprehensive Human Resource Development programme should be launched to train 1,000 young scientists in the research pertaining to solar oriented science and technology. One hundred scientists should be trained at M.Tech and PhD level in 'world class' universities and research institutions. On the whole 100,000 scientists and technologists are expected to work in the arena of solar energy by the end of the third phase.

A National Centre of excellence *viz.*, the Solar Energy Centre should coordinate and guide the research projects. Centres of excellence should be set up across India to enable specialisation in various solar technologies and scientific sub-branches. Engineering and management institutions and colleges led by IIMs and IITs should set up Solar Energy Department and conduct courses on the science and technology aspects of solar energy.

Future Thrust

In addition to the R&D projects supported by the MNRE and the private corporations, the programme would support R&D for the establishment of:

1. 50-100 MW solar thermal plant with 4-6 hours' storage (capable to meet both morning and evening peak load demands and double plant load factor by up to 40 per cent).
2. A 100-MW capacity parabolic trough technology-based solar thermal plant.
3. A 100-150 MW solar hybrid plant with coal, gas or biomass to overcome variability problem and space constraints.
4. 20-50 MW solar plants with/without storage, based on central receiver technology with molten salt/steam as the working fluid and other emerging technologies.
5. Grid-connected rooftops PV systems on selected government buildings and installations, with net metering.
6. Solar-based space cooling and refrigeration systems for selected government buildings and installations capable of meeting the peak load demands during the day time and during the summer season.

In concurrence with the development of science and technology, appropriate support should be given by the Centre of Excellence and IIMs in the setting up of manufacturing plants. The Centre for Innovation, Incubation and Enterpreneurship in IIM Ahmedabad and similar such centres should provide mentoring, guidance and venture capital support.

Ten per cent of returns to the venture capitalist *viz.*, the Government in this case should be invested in further support to R&D. Multinational corporations and foreign governments should be engaged in the transfer of latest technology and the provision of Intellectual Property Rights (IPRs) to the Indian manufacturers.

As the R&D transitions into application orientation from the development orientation, the Government should set up the 'open source' knowledge bank pertaining to solar power further contributing to the up-scaling and cost reduction leading to reliability and sustainability of power supply and energy security.

Monitoring and Guidance of the National Solar Mission

The National Solar Mission will be steered by a Mission Steering Group chaired by the Minister of Renewable Energy. Relevant stakeholders *viz.*, the representatives from various ministries will be the members. The group will approve new projects, technologies and manufacturing plants and funding for the projects. The Mission Executive Committee headed by the MNRE Secretary will oversee the implementation of the projects and monitor the performance of ongoing projects.

The Industry Advisory Committee headed by an eminent scientist with the rank of 'Joint Secretary' advises on the matters pertaining to the choice of technology and the level of technology absorption in the industry. The Mission Directorate monitors and evaluates the projects on a 'day-to-day' basis contributing to the time-bound achievement of goals and objectives. The directorate is headed by the Mission Secretary and comprises scientific officers ranked at 'Joint Secretary' level and other experts and consultants.

Wind Power Profile

Wind power is the dominant form of renewable energy in use. Almost three GW of power is derived from wind energy. About 5 per cent of the world's electricity is produced from wind with the Scandinavian countries leading the way with about 20 per cent soon to be achieved by the UK. The economies of scale pertaining to the land use have been augmented overtime and consequently, the mass production has become a possibility on a smaller area both on and offshore.

Wind Power Technology Development

United States and several European countries are leading the way in wind power technology development (GWEC, 2006). The technologies are of diverse range *viz.*, improving the efficiency of generators, enhancing blade pitch control diminishing generator noise, methods for modelling, verifying the efficiency of generators at low or middle wind velocity and air current changes around generator parts.

Other areas of research include the novel turbine, generator and systems designs aimed at reducing the unit costs and intelligent control systems to optimise the operation of turbines. The wind power technologies are beset with certain challenges. Rotor noise, visual impact and the risk of collision between the rotor blades and flying objects such as birds are some of the important challenges to the wind power technology.

The International Energy Agency has coordinated research for a number of offshore and large-scale wind power projects under the so-called 'Wind Task'. The United States is involved in major offshore and onshore wind power generation projects. The projects are led by the National Renewable Energy Laboratory (NREL) and Sandia National Laboratories. Denmark intends to install wind power plants with a capacity of 5,500 MW (4,000 MW offshore) by 2030 under the 'Energy 21'—The Danish government's primary plan for new energy development.

In China and India, the wind power industry has developed quickly. Burgeoning energy needs and strong economic growth have contributed to the rapid growth in renewable: especially wind energy. India ranks fifth with a total wind power generation capacity of 12,178 MW. (GWEC, 2010 and MNRE, 2010). China overtook India in 2008 and now in 2010 has a capacity of over 15,000 MW.

Offshore Wind Power Generation

Research into offshore wind power generation has been driven primarily by the shortcomings of onshore wind power production. Compared to the onshore, offshore wind turbines benefit from higher wind velocities and smaller variations in wind direction. Hence, larger-scale installations can be established and spatial limitations are substantially mitigated.

Denmark, Netherlands, Germany and the United Kingdom are developing offshore wind power generation facilities with turbines installed at a depth of between 5 and 20 metres and at a distance of several kilometres from the shore. USA and China are the new entrants. The scale economies to be derived from the offshore installations would substantially reduce the unit costs if the cluster or a wind farm is able to generate several hundred MWh of power and securely transmit to onshore.

Concluding Remarks

Solar PV applications and solar thermal applications are being given greater impetus in India. Unlike the present domination of the off-grid applications, the paramount is given to grid-based applications by the JNNSM. Solar collector capacity is growing 'rapidly in the world and in leaps and bounds' in India *viz.*, at an annual rate of almost 25 per cent. However, India is constrained by the low temperature, low efficiency solar PV modules.

Hence, JNNSM gives impetus to R&D and manufacturing of high and medium temperature applications through refinance, tax incentives, renewable purchase obligations and through incubation support. Provision of IPRs through international cooperation and collaboration and the gradual setting up of 'open source' knowledge bank pertaining to solar power are the highlights.

Since India's solar PV sector is dominated by 15 companies producing only up to 700 MW of modules, impetus is given for small and medium enterprises to produce and market taking the total capacity up to 8 to 10 GW by the year 2022 *viz.*, by the end of the third phase of JNNSM. From the present 10 kW limit of SPV power plant, as envisaged by MNRE, SRC is envisaging R&D projects to set up plants up to the capacity of 100 MW and hybrid plants *viz.*, solar/wind and solar/coal of up to 150 MW respectively. For this purpose, research and technological base is necessary. Premier educational institutions *viz.*, the IITs and the IIMs are required to take the lead in the technological and managerial education pertaining to solar technology and management. Up to 100 world class scientists and technologists, up to 1,000 good scientists and up to 1,00,000 technicians are expected to work in the solar energy sector in India by the year 2020. Appropriate monitoring and implementation mechanisms are already set in place by the JNNSM under the eages of the Mission Steering Group, the Mission Executive Committee, the Mission Directorate and the Industrial Advisory Committee respectively.

Wind power has become the dominant source of renewable energy. However, it will be replaced by solar power as the potential of wind power is limited. However, it is essential to note that over 5 per cent of the world's electricity requirements are met by the wind power. In India, the figure has already reached almost 8 per cent and the potential would be 48,000 MWh or 6 per cent of India's energy requirement in 2030.

11 Enhancing the Deliverables of the Power Sector

Introduction

As we discussed in Chapter 6, power sector is the key driver for India's economy. Even when we look at the current meagre requirements of the present times *viz.*, over 700 kW per capita, India's power generation falls short by about 11 per cent during off peak and 15 per cent during the peak time. This implies that India is generating just over 1,70,000 MW of power *vis-à-vis* the current requirement of over 1,90,000 MW. Compare this with the 9,50,000 MW requirement in the year 2031. While the population of India is expected to be 25 per cent higher, the per capita power consumption is supposed to be four-fold, or about 3,000 kW providing a comfortable living standards.

The year 2020 itself requires the more than doubling of India's power generation capacity to about 4,00,000 MW. In the Twelfth Five Year Plan alone, about 1,00,000 MW of capacity needs to be added with the capital expenditure of about Rs 7 lakh crore for power generation, and about Rs 5 lakh crore for establishing the transmission and distribution infrastructure, upgradation of the human resources, equipment manufacturing capability etc. A comprehensive project planning and execution capability, programme formulation, implementation, monitoring and evaluation mechanism, and delivery of the deliverables assume very important role. This chapter has seven sections.

Section II discusses the project management principles in general. Section III discusses the power generation capacity and transmission capacity of India, and about the institutions involved in promoting energy solutions, efficiency, rural electrification and renewable energy. Section IV discusses the Ministry of Power's strategy blueprint in relation to the project/programme management principles. Section V discusses the Ministry of New and Renewable Energy strategy blueprint in relation to the project/Programme management principles. Section VI discusses the evaluation principles used by the Comptroler and Auditor General of India using the NTPC (The National Power Company) as the case in point. Section VII presents the concluding remarks.

Principles of Project Management

The Background

The manufacturing, supply and the servicing of a project usually takes place in a timeframe. If the aforesaid activities occur within a predictable timeframe, the cost and revenue structures are more likely to be predictable. If there is any deviation in the timeframe of any of the steps/ processes, the cost and revenue structures will be affected and would be severely impacted upon. In some cases, the failure of the project to deliver would make an entire company or even the corporation insolvent.

The delivery of a particular product or a service takes place through a venture. The venture may be a company in itself or may be a component of a company. Such a venture is also called 'project'. Human beings functioning as managers should first learn to manage a venture before managing the entire company, then corporation, then a province, then the country and finally the United Nations Organisation as a whole.

What is a Project?

A project is a unique venture starting and ending on pre-specified dates. Unlike the project, the ongoing task does not seem to have a pre-specified end date. Projects are run by teams of people representing and involving different parts of an organisation.

Constraints on the Project

There are four main constraints on the initiation and implementation of a project.

a. Cost,
b. Schedule,

c. Resources and

d. Quality.

Often, these constraints are not complimentary. Attempts to overcome one constraint results in greater difficulty to overcome the others. This is known as the trade-off. For example, attempts to reduce the time schedule would lead to increased cost on account of relatively scarce resources. This is the typical problem of the short run economies of scale. Similarly, poor quality of the resources would reduce the quality of the final product. If the schedule is also hastened, along with the reduction in the quality, costs would also increase resulting in resource constraints. To overcome the adverse effects of the trade-off, a project is divided into segments and also stages or phases. Each phase has its own set of priorities and goals.

What is Project Management?

'Project management is a combination of techniques, procedures, people, and systems focused on the successful completion of a project'. It is also a discipline supporting the planning, implementation, tracking and control of projects.

A project, or a few projects may not meet the policy objectives of a country or a corporation or a company. Hence, a programme is designed and implemented towards meeting the policy objectives.

What is a Programme?

A group of interrelated projects are called a programme. These projects are often mutually dependent and the relationship tend to be interwoven in a complex manner. A coordinated management of the projects is the key to the success of the programme.

Typical Project Phases

a. Definition,

b. Planning,

c. Procedural approvals,

d. Implementation and

e. Evaluation.

Understanding and Managing the Project Politics

Projects are rife with politics. Every stakeholder tends to jockey to portray the importance of, and to achieve the goals they deem to be most important. Project manager and the team need to manage the politics on a priority basis. Project manager needs to help the team members to overcome, or to circumvent the political situations. Information sharing is the key to managing the politics. Larger project makes it time consuming for the information to be shared among the interested parties.

Balancing the Priorities

A project comprise a set of priorities. The initial goals, the re-definition of the goals subject to the constraints, and the changing circumstances, and the impact on the choices of the team members, all have to be understood, accomplished, and balanced by the project manager.

Scope of the Project

Criteria for the project success, and the actual accomplishment should be carried out to the optimum satisfaction of the stakeholders *viz.*, the investors, the customers, the environmental groups, the governments, the third parties with perceived externality effects etc. Timely verbal, pictorial, graphic and focus group communications, brainstorming discussions etc., will help achieve the goals. Besides the stakeholder's satisfaction, this process enables the capturing of the requirements. Emphasis should be laid on clarity rather than on the formal methods of communication and documentation.

This process helps in the assessment of the feasibility, prior to the actual commencement of the project. Technical, organisational and financial feasibility and support structures should be analysed in the feasibility report. Business case analysis and cost-benefit analysis will help the process. The outcome will decide the going ahead, or the shelving of the project.

Project Risks

Risks eminate from uncertainties and assumptions at the outset, and their dynamics during the progress of the project.

At the outset, the list of risks should be reviewed, and their relationship with the project stages should be well understood.

Project Objectives: A Presumptive Risk Mitigation Strategy

The project objectives are a critical guide posts for the project. They define the success parameters, and achievements by the project manager, and the goals for the project associates. The document also reflects the contracts with the customers, and how they can be tailor-made to suit their changing needs and new requests. The objectives should contain the goals statement and the list of deliverables. Participation of the stakeholders, and the hearing of all of their views is the prior responsibility of the Project manager before crafting the document and collecting the signatures. In addition to the smoothening of the project performance, this will also contribute to the preemptive avoidance of the legal problems which in themselves would lead to a significant financial saving for the company/ corporation, besides, to the customers, government, and to other relevant stakeholders.

Statement of the Project Goals

This 50-word statement discusses the purpose, conditions, implementation timeframe, and constraints of the projects.

Defining Project Deliverables

Deliverables are tangible and measurable. External project stakeholders are going to receive them when the project is complete, or at the completion of each phase. Deliverables include: equipment, report, installation of new hardware, etc. At the outset, work, or product, or organisation, or the cost breakdown structure should be provided to the stakeholders for the purpose of establishing clarity. This helps establish the accuracy of the cost estimates, the timeframe of the work completion for a person, or for a group etc.

Importance of Good Estimates

Good estimates help in the expedition of the approval, and further approval for the continued work. Stakeholders usually concern about the rate of return, and the unforeseen costs and risks. Benchmarking is the key and a relatively easier way for providing good estimates. Estimates should be prepared by taking right amount of time with right person. It is not advisable to haggle over estimates with right kind of person. In the process of building trust, estimation range is better than the arbitrary citing and quotation of figures. Comparing costs and improving upon the estimations is part of the project management process. Hence, fuzzy large estimates should be justified by breaking down the costs.

Good estimates can be arrived at by sequencing of the tasks wherever dependencies are inevitable, and parallel activities can be undertaken wherever not. Estimates should capture the skill set of the team members in a disaggregated manner, taking into account the differential experience level and multifacetedness of the skills.

Having established sequencing, the 'critical path' should be established involving a network of tasks to accurately reflect the time span for the completion of individual tasks, groups of tasks, and the project as a whole.

Resource Justification

Project managers usually bank on the shared pool of resources. Proper justification should be given while requesting for additional resources, or while transferring resources from one task to another. Some tasks might overrun the expenditure, and some might involve lesser expenditure.

The Importance of Good Team

The success of a project hinges on the building of a good team. A good team provides good deliverables that are almost flawless and error free. Good team is vibrant, dynamic, and replete with new ideas and implementation methods. The following methods enable the building of a good team.

A. *Openness and honesty*: Project manager, and other responsible people should serve as a conduits of information. If something can't be disclosed, it should be stated frankly so that, the rest of the team will involve more in work and less in gossip.

B. *Equality and fairness*: This promotes merit-based judgement, and hence enable the arrival at accuracy.

C. *Team ownership of the problem and its solution:* Let the team shoulder the relevant problem, discuss, and find the ways and means of solving them.

D. *Loyalty to the team:* If there is a conflict of interest between the team and client's satisfaction, inform the client, then discuss with the team and come up with the solution.

E. *Learn to delegate:* Trust the team leaders, delegate the right kind of tasks, and get the results, which is akin to the military jargon. Reasonable and achievable goals are thus given to the team leaders, so that the project manager concentrates on the provision of appropriate support. This procedure will provide sufficient autonomy to the teams, so that they will concentrate on their job, rather than disturbing the project manager impeding the undertaking of higher level tasks.

Manager's Primary Responsibilities

a. Tracking

Project manager should ensure the timely completion of tasks, and the project, while at the same time should ensure the quality, budgetary, and other forms of compliance. Issues of risks, challenges and contention should be tracked early and resolved. Tracking the factors and agents that are important for the success of the project is very important. Things should be known based on the need and not on the nicety. Tracking should be milestone based, rather than percentage based. Percentage basis is misleading and only leads to the falsification of the information. Paradoxically, it would take as much, or more time to complete the last 10 per cent of the job *vis-à-vis* the reported first 90 per cent of the job.

b. Control

Change is constant. Request for change in the aftermath of the commencement of the project might either lead to time, or cost escalation, or both of them. However, the assessment should be in relation to the impact. Thereafter, information should be shared with the stakeholders. This helps the project manager to continue to exercise control over the project, and the project is less likely to be derailed, and shelved. There is no trade-off between the trust and the status report in reality. Status report is only meant to keep the stakeholders informed of things and has got nothing to do with trust, or mistrust. Status reports either place the new items on the top, or the issues on the top, placing the closed items at the end. The frequency, the purpose, and the nature of the status report should be predictable and should not tax the project manager's time away from the actual developmental tasks.

c. Reflection

The time of the project closure should reflect upon what went wrong, why it went wrong, and find the ways of avoiding the bumps in the future projects. Finger pointing should be avoided as far as possible, and if possible wholly. Keeping the groups for meeting small, allowing the reflection after the closure and winding up of the projects, manager's courage to point her/his own mistakes, etc., contribute to the successful holding of the meetings, in a friendly manner. At the outset, the manager, or the chair of the meeting should begin the meeting with her/his own thoughts. Stakeholder inclusion to get the full range of perspectives and the organising of celebration to mark the project closure will go a long way to contribute towards goodwill and solidarity.

Axioms for the Project Success

a. Know your goal and state it in a single clear sentence.
b. Know your team, build trust, establish channels of communication, and establish merit-based reward structure.
c. Know your stakeholders to obtain commercial, technical and political expert advise. Establish cordiality, and grease the wheels whenever necessary in an appropriate manner. This will ensure smooth project ride.
d. Planning and design should lead to efficient and elegant solution of the problem, and burden sharing.
e. Promise low and deliver high, so that, personal, team, and the stakeholder confidence and trust gets established. This will also ensure contingency against things going wrong.

f. Generate positive atmosphere and receptivity. This will ensure recognition of success, and even recognition of reasons for failure that would not be pinpointed on the manager, or the team.

g. Iterate, increment and evolve. This will enable the breaking down of the tasks, provision of accurate estimates, and the identification and solution of the problems and contentious issues.

h. Stay on track: Budget, time schedule and the client's needs should all be tracked. This becomes more important if your ideas and work determines the very course of business.

i. Manage change: Evolving stakeholder needs, positive ideas and boreholes of the team members, and other detractory stakeholders, should all be controlled and managed to ensure timely and proper completion of the project. To avoid pitfalls, and the shattering of the project, planning, regular updation of the plan and the ensuring of the following of the plan should be ensured.

j. Test early, test often: Creative mistakes and assumptions can lead to fatal mistakes in the end products. Hence, testing early, and often will help avoid these pitfalls.

Ultimately, the project manager should keep an open mind concerning the methodologies and ideas. Without ideological barriers, the project manager should focus on the quality of the end product, its delivery and its promotion.

Then, celebration should follow and should move on to the next project.

India's Power Generation Requirements

By March 2011, India's power generation capacity reached about 170,000 MW. This is still about 11-15 per cent short of the capacity requirement.

India's power sector has high growth potential until the year 2030. India needs to add about 780,000 MW to reach the capacity to 950,000 MW in the year 2030.

However, India's primary planning regime encompasses a five-year perspective. The strategy document envisages a growth of about 130,000

MW of power generation capacity between 2011 and 2017 to reach the cumulative capacity to 300,000 MW.

Capacity Targets

India has always been falling short of the power generation capacity targets. While the target for the Tenth Plan was about 40,000 MW, the actual growth was only about 28,000 MW taking the figure to 138,000 MW. Eleventh Plan continues to suffer from the shortfall of the target. Half of it is attributed to a shortfall in rainfall, and shortage of quality coal by the Ministry of Power in its Annual Report 2009-10.

Capacity Addition Programme in the Eleventh Plan

The National Electricity Policy (NEP) projects that the annual per capita consumption of power will rise to 1,000 units by 2012. This entails the provision of adequate and reliable power at affordable price and access to all citizens. Electricity is in the Concurrent List of the Constitution and the primary responsibility of structuring its availability and distribution is that of the states. However, both the Centre and the states have to play a decisive and effective role.

To fulfill the objectives of the NEP 2006, a capacity addition of 78,700 MW has been proposed for the Eleventh Plan. This capacity addition is expected to result in a compound annual growth rate (CAGRA) of 9.5 per cent. The breakup of the capacity addition target is given as under:

Table 11.1

Capacity Addition Targets in the Eleventh Plan

(in MW)

Source	*Central*	*State*	*Private*	*Total*	*Share (%)*
Hydro	8654	3482	3491	15627	19.9
Thermal	24840	23301	11552	59693	75.8
Nuclear	3380	-	-	3380	4.3
Total	36874	26783	15043	78700	100
Share (%)	46.9	34	19.1	100	

India has to realise enormous amount of hydropower potential. The unrealised potential stands at over 80,000 MW, of which, projects are allotted to realise over 58,000 MW. Most of the allotted projects are in the northeastern region of India.

Besides, the entire hydropower potential of Bhutan ranging up to about 80,000 MW is at India's disposal, given the special relationship.

In states like Himachal Pradesh, Arunachal Pradesh, Andhra Pradesh, Chhattisgarh etc., small hydro projects have good potential.

The Steering Committee of the Eleventh Plan has approved 1,344 small hydropower projects under Part-A at the cost of Rs 4859.61 crore for 22 states

Table 11.2

Approved Small Hydro Projects under the Eleventh Plan

Sl. No.	*State*	*No. of Projects Sanctioned*	*Sanctioned Project Cost (Rs Crore)*
1.	Andhra Pradesh	113	388.02
2.	Bihar	71	194.6
3.	Chhattisgarh	20	122.45
4.	Goa	4	110.72
5.	Gujarat	84	225.36
6.	Haryana	36	165.63
7.	Himachal Pradesh	14	81.07
8.	Jharkhand	30	160.61
9.	Karnataka	98	391.14
10.	Madhya Pradesh	82	228.09
11.	Maharashtra	130	252.11
12.	Punjab	47	272.85
13.	Rajasthan	87	315.93
14.	Sikkim	2	26.3
15.	Tamil Nadu	110	417
16.	Uttarakhand	31	125.82
17.	Uttar Pradesh	168	665.5
18.	West Bengal	62	159.98
19.	Assam	66	173.18
20	Kerala	43	214.4
21.	Tripura	16	34.36
22.	Jammu & Kashmir	30	134.49
	Total	1344	4859.61

The below tables discuss the Central and state sector hydropower projects completed in the Tenth Plan, and those to be completed in the Eleventh and Twelfth Plans, along with their capacity targets and estimated cost.

Table 11.3

Hydro R&M Schemes Completed upto Tenth Plan

Sl. No.	Plan Period	No. of Projects			Installed Capacity (MW)	Estimated Cost (Rs in Crore)	Actual Expenditure (Rs in Crore)	Benefit (MW)
		Central Sector	State Sector	Total				
1.	Up to Eighth Plan schemes completed	2	11	13	1282	126	127	429
2.	Ninth Plan Schemes completed	8	12	10	4892	598	570	1093
3.	Tenth Plan Schemes completed	5	27	32	4337	1016	1029	829

Table 11.4

Hydro R&M Schemes Planned during Eleventh Plan

Sl. No.	Plan Period	No. of Projects			Installed Capacity (MW)	Estimated Cost (Rs in Crore)	Actual Expenditure (Rs in Crore)	Benefit (MW)
		Central Sector	State Sector	Total				
a)	Programmed (revised)	10	46	56	10810	4203	842	4099
b)	Completed	2	6	8	950	228	146	380
c)	Ongoing	5	21	26	7575	1757	685	2195
d)	Under tendering	2	6	8	1341	1065	10	453
e)	Under DPR finalisation	1	7	8	500	457	1.00	506
f)	Under RLA studies	-	6	6	445	695	-	565

Note: DPR: Detailed Project Report; RLA: Residual Life Assessment.

Table 11.5

Hydro R&M Schemes Programmed for Twelfth Plan

Sl. No.	Plan Period	No. of Projects			Installed Capacity (MW)	Estimated Cost (Rs in Crore)	Actual Expenditure (Rs in Crore)	Benefit (MW)
		Central Sector	State Sector	Total				
a)	Programmed	-	22	22	2467	1066	0.70	2246

Given the special hydropower potential of the northeastern states, especially Arunachal Pradesh and Sikkim, and great strategic importance of this region for India, the Eleventh and Twelfth Plans have allotted large number of projects. The total allotment for the period including the period beyond Twelfth Plan is to the tune of over 57,000 MW in this region.

The tables below discuss the situation.

Table 11.6

Hydropower Projects Allotted for the Northeastern Region for the Twelfth Plan

State	*No. of Schemes*	*Hydro(MW)*
Arunachal Pradesh	19	7803
Assam	1	150
Manipur	1	66
Meghalaya	0	0
Mizoram	0	0
Nagaland	0	0
Tripura	0	0
Total (NER)	21	8019
Sikkim	1	99

Table 11.7

Hydropower Projects Allotted in Northeastern Region

Name of State	*Central*		*Private*		*Unallocated/State*		*Total*	
	No.	*IC (MW)*	*No.*	*IC (MW)*	*No.*	*IC (MW)*	*No.*	*IC (MW)*
Arunachal Pradesh	4	5870	74	27512	8	14785	86	48167
Assam	-	-	-	-	5	360	5	360
Manipur	2	1566	-	-	3	195	5	1761
Meghalaya	1	90	7	1745	7	635	15	240
Mizoram	3	730	0	0	1	60	4	790
Nagaland	-	-	-	-	1	140	1	140
Total (NER)	10	8256	81	29257	25	16175	116	51458
Sikkim	2	730	22	2577	1	90	25	3397
Total (NER + Sikkim)	12	8986	103	31834	26	16265	141	54855

Note: IC: Installed Capacity.

Besides the targetted capacity addition, the policy objective of providing quality and adequate power is possible through rural electrification.

As mentioned and discussed in the earlier chapters pertaining to the renewable energy, the National Energy Policy of 2006 aims at electrifying all the households in the country, and providing a daily household electricity consumption of at least one unit by the year 2012. The revised target fixes the deadline at the year 2018.

Table 11.8 shows the progress of village electrification until the year 2010, and Table 11.9 shows the progress of rural electrification under the Rajiv Gandhi Grameen Vidyutikaran Yojana (RGGVY) launched in the year 2006.

Table 11.8

Year-wise Village Electrification

Sl.No.	*Years*	*No. of Villages Electrified*	*Remarks*
1.	1990-91	10286	
2.	1991-92	6046	
3.	1992-93	3669	
4.	1993-94	3352	
5.	1994-95	3554	
6.	1995-96	4086	
7.	1996-97	3843	
	Eighth Plan (Total)	*34836*	
8.	1997-98	3207	
9.	1998-99	2780	
10.	1999-00	2093	
11.	2000-01	1218	
12.	2001-02	4118	
	Ninth Plan (Total)	*13416*	
13.	2002-03	2626	
14.	2003-04	3352	
15.	2004-05	3884	
16.	2005-06	9819	RGGVY
17.	2006-07	28706	RGGVY
	Tenth Plan (Total)	*48387*	
18.	2007-08	9301	RGGVY
19.	2008-09	12056	RGGVY
20.	2009-10 (upto 31-1-2010)	11911	RGGVY

Source: Ministry of Power. *http://www.powermin.nic.in/.../Achievement_of_Rural_Electrification_ Targets.pdf*

Table 11.9

Cumulative Performance under RGGVY

Tenth Plan	235	68763	111936	83.1 lakh	9732.9
Eleventh Plan	332	49736	242733	162.96 lakh	16506.04

Source: Ministry of Power. *Outcome Budget 2011-12*.

The above discussion suggests that the responsibility for power generation is that of the enterprise. But the responsibility for power distribution is a entrepreneurial cum social responsibility. Central government tends to involve in power generation that bestows profit. Whereas, the state government needs to pick up the subsidy bill. Thus, the revenue generation capability of the project managers related to the state governments is curtailed at the outset.

Similarly, the extent of revenue generation capability determines the extent of modernisation, upgradation, and the quality of service delivery capability. The very progressive rate of power tariff, along with subsidy to farmers and other vested interests, such as the new and upcoming enterprises causes a dent in the revenue, service capability, and the capability to execute fresh projects.

The major question that arises, which requires urgent resolution is whether the project managers in India: especially in the power distribution sector work in a tight budget, tight constraint regime?

There is a need for a comprehensive project report in this respect. Legitimate constraints need to be identified, and redressed to the best possible extent. There is a need for reviewing the credit structures and payment durations. The ultra modern power lines for transmission and distribution would service the nation for more than half a century. Besides, they would be laid underground and would be tamper-proof and theft-proof. Right kind of project report would be friendly to the cause of project managers, even though they work under a relatively tight revenue regime.

Power Shortage

Section VII in Chapter 3 discussed about the demand/supply gap in the power sector in the country as a whole, as well as across the states. In this chapter, we discuss about the shortage in the context of the targetted capacity addition.

Contrary to the present optimistic expectation about a CAGRA of 9.5 per cent for the Eleventh and Twelfth plan periods, the past performance in the growth of the power sector is dismal. In the 13-year period *viz.*, between 1997 and 2010, the sector grew only by about 54 per cent, or by about 4 per cent per year. The power generation capacity grew from 420.5 billion units to about 650 billion units between the financial years, 1997-1998 and 2009-10. In fact, almost 8 per cent of the growth, or about 50 billion units of growth is attributable to the growth in the plant load factor (PLF): especially in the thermal power plants.

The PLF in the thermal power plants grew from 71.1 per cent to 76.65 per cent in the corresponding period. In addition, the still higher transmission and distribution losses of over 30 per cent in the country cause a dent in the revenue generation, modernisation, upgradation and capacity addition efforts. Even during this Eleventh Five Year Plan, capacity addition is expected to fall short by about a third of the target. In other words, India is only expected to generate 190,000 MW of power. As against the total generation of about 3,380 MW of nuclear power in the Eleventh Plan, only about 600 MW has been added during the first four years. Even if another 500 MW of capacity gets added, the shortfall is still by about 2/3rd.

Based on the above discussion, we understand that the demand/supply gap persists now, and would persist even until the year 2017. Table 11.10 and 11.11 capture this phenomenon.

Table 11.10

Comparative Projections: 17th EPS, IEP versus Actual

	Peak Demand (MW)		*Energy Requirement (BU)*	
	2011-12	*2016-17*	*2011-12*	*2016-17*
17th EPS	1,52,746	2,18,209	969	1,392
IEP (8%)	1,21,000	1,87,000	895	1,243
Projected based on actual	1,36,249	1,90,451	954	1,348

Note: EPS: Electric Power Survey; IEP: Integrated Energy Policy.

Table 11.11

All-India Power Deficit (Energy Units and Electricity)

Sector	*Requirement*	*Availability*	*Deficit*	*Deficit (%)*
2009-10				
Energy (BU)	830.5	746.6	83.9	10.1
Peak electricity demand (MW)	119,166	104,009	15,157	12.7
2010-11 (Apr-Dec)				
Energy (BU)	638	582	56	8.8
Peak electricity demand (MW)	119,437	107,286	12,151	10.2

Source: Ministry of Power. Strategic Blueprint.

Monitoring Mechanism to Track the Capacity Addition

The Ministry of Power has adopted a robust monitoring system for the capacity addition programme so as to see that the cleared projects are executed in time. The monitoring mechanism comprises three broad levels at which monitoring of power projects are carried by the Ministry *viz.*, by the Central Electricity Authority (CEA); and Ministry of Power through the Power Project Monitoring Panel (PPMP).

The Central Electricity Authority (CEA) has a nodal officer associated with each ongoing project which continuously monitors the progress at site through frequent visits and continuous interaction. The respective nodal officer is responsible for submitting a report on the progress of each of the ongoing power project on monthly basis highlighting the critical areas where corrective actions are required. The chairperson, CEA reviews the progress every quarter.

Intensive reviews are held by the Ministry of Power to review the critical milestones associated with each ongoing project. Meetings with the leading equipment manufacturers, especially BHEL are organised to review the critical supplies to the projects. Quarterly performance reviews (QPRs) are also organised separately for each CPSU to review the status of the Central sector projects. An Advisory Group under the chairmanship of the Hon'ble Minister of Power has been set up to suggest the ways and means to achieve the Eleventh Plan capacity addition target. The Eleventh Plan Capacity Addition Programme is also monitored by the Planning Commission, PMO and the Cabinet Secretariat as well.

The Ministry of Power has set up a 'Power Project Monitoring Panel' (PPMP) for monitoring of thermal and hydro generation projects targetted for commissioning during the Eleventh Plan along with the associated transmission schemes. The PPMP at present comprises five independent project monitoring consultants. Each consultant is given specific projects. The individual consultants make visits to the project sites and furnish their progress report which is compiled by the coordinating consultant and along with the Exception Report submitted to the Secretary. The progress of implementation of the projects is accordingly reviewed by the Ministry on the basis of the report received from the monitoring panel.

Accelerated Power Development and Reforms Programme: A Road towards a Modern and Efficient Power Distribution System

The Scheme

The Cabinet Committee on Economic Affairs (CCEA) approved the 'Re-structured APDRP' for Eleventh Plan as a Central government scheme in its meeting held on 31.07.2008. The focus of the programme is on actual, demonstrable performance in terms of aggregate technical and commercial (AT&C) loss reduction. Projects under the scheme would be taken up in two parts in urban areas—towns and cities with population of more than 30,000 (10,000 in case of special category states). Projects execution under the scheme are to be taken up in two parts. Part-A shall include the projects for establishment of baseline data and IT applications for energy accounting/auditing & IT-based consumer service centres. Part-B shall include regular distribution strengthening projects. The activities to be covered under each part are as follows:

- Part-A: Preparation of baseline data for the project area covering consumer indexing, GIS mapping, metering of distribution transformers and feeders and automatic data logging for all distribution transformers and feeders and supervisory control and data acquisition (SCADA)/distribution management systems (DMS) for big cities only. It would include asset mapping of the entire distribution network at and below the 11 kV transformers and include the distribution transformers and feeders, low tension lines, poles and other distribution network equipment. It will also include adoption of IT applications for meter reading, billing & collection, energy accounting & auditing, redressal of consumer grievances, establishment of IT enabled consumer service centres etc. The baseline data shall be verified by an independent agency appointed by the Ministry of Power.

- Part-B: Renovation, modernisation and strengthening of 11 kV level.
- Substations, transformers/transformer centres, re-conductioning of lines at 11 kV level and below, load bifurcation, load balancing, service centres etc. In exceptional cases, where subtransmission system is weak, strengthening at 33 kV or 66 kV levels may also be considered.

Expected investment in Part-A (baseline system) shall be Rs 10,000 crore and that in Part-B shall be Rs 40,000 crore.

Initially 100 per cent funds for Part-A and 25 per cent (90 per cent for special category states) funds for Part-B projects shall be provided through loan from the Government of India. The balance funds for Part-B projects shall be raised from financial institutions.

The entire amount of loan for Part-A projects shall be converted into grant once the establishment of the required baseline data system is achieved and verified by an independent agency appointed by MoP.

Upto 50 per cent (90 per cent for special category states) of the project cost of Part-B projects shall be converted into grant in five equal tranches on achieving the 15 per cent AT&C loss in the project area on a sustainable basis for a period of five years. In addition, utility level loss reduction (AT&C losses) @ 3 per cent per annum for utilities with baseline loss levels exceeding 30 per cent and @ 1.5 per cent for utilities with baseline loss levels less than 30 per cent have to be achieved.

Part-C of the programme is an enabling component for the implementation of Accelerated Power Development & Reforms Programme (APDRP). Provision of Rs 1,177 crore through gross budgetary support (GBS) has been provided in the scheme. This part is to be implemented by Ministry of Power/nodal agency. Power Finance Corporation (PFC) has been appointed as nodal agency for operationalising the programme. The following activities are included in Part-C:

Preparation of a template for system requirement specifications for subdivision automation and for customer relations management module, as well as for baseline data collection, agencies will also verify the AT&C losses and monitor quality of works to be executed under Part-B.

Project advisors and project management consultants: Advisor cum consultants will be appointed to assist the Ministry in monitoring of APDRP and to validate the project proposals submitted by the distribution companies. Project management consultants will assist distribution companies in formulating the DPRs, in standardisation of bidding/contract documents, managing the bid process, monitoring of progress, quality assurance etc. They will also facilitate the management information system and assist the Distribution Reforms Committees formed at the state level. Project evaluation by third party will be the basis of computation of the extent of conversion of loan into grant for the specific project. A panel of project evaluators will be finalised through a bidding process.

Capacity building and development of franchises in distribution sector will be a major focus area to provide training to employees of the distribution companies and existing and prospective franchises in management, technical, commercial and consumer related areas, exposure to latest developments in electricity distribution, loss reduction, theft and pilferage control within India and abroad, dissemination of knowledge through best practice workshops and conferences, standardisation of specifications of equipment required in electricity distribution network, standardisation of contractual documents for outsourcing project management, turnkey jobs, franchising etc.

Consumer Attitude Survey will be carried out to assess the impact of the measures taken in the distribution sector towards improving of service.

Under Part-D of the scheme, there is provision for incentive for utility staff in towns where AT&C loss levels are brought below the baseline levels *viz.*, if brought below 15 per cent.

Capacity Building under R-APDRP

In order to reap the benefits of the R-APDRP programme, the capacity building of power utility personnel at various levels has been identified as an essential part of R-APDRP. This portion has been extensively covered under Part-C of R-APDRP's guidelines with focus on enhancing skills of utility personnel at various levels for efficient management and operation. The Ministry has also constituted an Advisory Committee for R-APDRP capacity building to advise PFC for developing a comprehensive capacity building plan for employees of distribution utilities under R-APDRP. PFC is in the process of empanelling partner training institutions and resource institutes which shall be associated in delivering training to the utility personnel.

Implementation of R-APDRP

A Steering Committee under Secretary (Power) comprising representatives of Ministry of Finance, Planning Commission, Central Electricity Authority, Power Finance Corporation, Rural Electrification Corporation, selected state governments (on one year rotation basis) and of Ministry of Power has been constituted. The Steering Committee will sanction projects, including modification or revision of estimates; monitor and review the implementation of the scheme; approve the guidelines for operationalisation of various components of the scheme including the approval of the charges to be paid to the nodal agency; approve and sanction activities to be taken up by the Ministry under Part-C of the scheme; appoint agencies for verification and validation of baseline data systems, for verifying the fulfillment of programme conditions by utilities and approve conversion of loan into grant upon fulfillment of the necessary conditions.

Strategy Blueprint of the Ministry of Power

Demand Assessment

India's electricity additional requirement is assessed by the 17th Electric Power Survey at 190,451 MW in 2017 as compared to 2007. The survey has accounted for a growth rate in demand as well as a growth rate in the requirement for power. The total electricity generation is supposed to be about 329,000 MW in 2017.

In the case of India, the requirement is more than the demand by about 10 per cent. The shortfall in the growth rate of the requirement for power by about 2.5 per cent compared to the actual estimates has revised the additional requirement projection downward from about 214,500 MW to about 190,451 MW.

In reality, only about 75,000 MW of power is likely to be added in the Twelfth Plan, or between the years 2012-2017.

Coal is still likely to contribute about 64 per cent, or about 48,000 MW, followed by gas, hydro and nuclear power respectively.

The below tables illustrate this fact.

Table 11.12

Capacity Addition Plan (Generation) for 2012-2017 (MW)

Sector	*2012-13*	*2013-14*	*2014-15*	*2015-16*	*2016-17*	*Total*
Capacity addition (MW)	17,500	16,000	15,500	14,000	12,000	75,000

Source: Same as Table 11.11.

Table 11.13

Capacity Addition Plan (Generation) for Twelfth Five Year Plan (MW)

Sector	*Share (%)*	*Total(MW)*
State	17	12,750
Private	54	40,500
Centre	29	21,750
Total	100	75,000

Source: Same as Table 11.11.

Table 11.14

Capacity Addition by Fuel

Sector	*Capacity Addition(MW)*
Hydro	10,150
Nuclear	3,800
Gas	13,000
Coal	48,050
Total	75,000

Source: Same as Table 11.11.

Assessment of Supply Position

Along with the demand, an assessment and understanding of the supply position and growth rate is also necessary.

The supply availability depends upon the installed capacity, operational efficiency, plant utilisation matrix and the system reliability matrix.

Operational efficiency is improving with the improvement in the PLF. PLF already increased from about 64 per cent in the thermal power plant to about 77 per cent between 1997 and 2010. This will further increase to beyond 90 per cent with the adoption of super critical and ultra super critical technology. Twenty-one thermal power plants experienced over 90 per cent PLF in 2010.

Similarly, the available transfer capability (ATC), and the system reliability matrix account for the supply position through the transmission and distribution network. The system availability has increased to 99.7 per cent with the increased efforts towards the nationalisation of the power grid through the aegis of the Power Grid Corporation of India Ltd. (PGCIL). PGCIL, presently accounts for over 2/3rd of the total transmission capacity.

The following figure shows the actual and projected growth rate in the transmission capacity in all the kV segments.

Table 11.15

Existing Interregional Power Transfer Capacity (MW)

	Existing (Nov. 2009)
East-North	
Dehri-Sahupuri 220 kV S/C	130
Sasaram HVDC back-to-back	500
Muzaffarpur-Gorakhpur 400 kV D/C (with series compensation + TCSC)	2000
Patna-Balia 400 kV D/C (quad)	1600
Biharshariff-Balia 400 kV D/C (quad)	1600
Sasaram bypassing (additional capacity)	500
Subtotal	6330
East-West	
Budhipadar-Korba 220 kV 3 ckts.	390
Rourkela-Raipur 400 kV D/C (with series compensation +TCSC)	1400
Ranchi-Sipat 400 kV D/C (with series compensation)	1200
Subtotal	2990

contd...

...contd...

	Existing (Nov. 2009)
West-North	
Vindhyachal HVDC back-to-back	500
Auraiya-Malanpur 220 kV D/C	260
Kota-Ujjain 220 kV D/C	260
Gwalior-Agra 765 kV D/C	1100
Gwalior-Agra 765 kV D/C 2nd ckt(NR-WR interregional strength scheme)	1100
Zerda-Kankroli 400 kV D/C (NR-WR interregional strengthening scheme)	1100
Subtotal	3220
East-South	
Gajuwaka HVDC back-to-back	1000
Balimela-Upper Sileru 220 kV S/C	130
Talcher-Kolar HVDC bipole	2000
Upgradation of Talcher-Kolar HVDC bipole	500
Subtotal	3630
West-South	
Chandrapur HVDC back-to-back	1000
Kolhapur-Belgaum 220 kV D/C	260
Barsur-L.Sileru 220 kV HVDC monopole	200
Ponda-Nagajhari 220 kV D/C	260
Subtotal	1720
East-Northeast	
Malda-Bongaigaon 400 kV D/C	1000
Birpara-Salakati 220 kV D/C	260
Subtotal	1260
Various 132 kV interregional links	600
Total (cumulative)	20750

Source: Same as Table 11.11.

The below tables show the planned addition to the transmission capacity in the Eleventh and Twelfth Plans.

Table 11.16

Transmission Network: Capacity and Target Additions (Eleventh Plan)

Infrastructure Planned	*Unit*	*End of Tenth Plan*	*Eleventh Plan Target Addition*	*Current Capacity (Oct, 2010)*	*End of Eleventh Plan*
Transmission lines	Ckm	198,089	68,955	244,521	267,044
765 kV/HVDC	Ckm	75,722	40,186	101,184	115,908
400 kV	Ckm	114,629	24,396	131,230	139,025
Substation capacity	MVA	259,639	126,700	329,312	386,339
+/- 500 kV HVDC converter/BTB station	MW	8,200	3,000	8,700	11,200
765 kV	MVA	2,000	24,000	4,500	26,000
400 kV	MVA	92,942	41,300	121,967	134,242
220 kV	MVA	156,497	58,400	194,145	214,897
National grid	MW	14,000	18,650	22,400	32,650

Table 11.17

Transmission Network: Target Additions during Twelfth Plan

Infrastructure Planned	*Unit*	*End of Eleventh Plan*	*Twelfth Plan Additions*	*End of Twelfth Plan*
Transmission (765–220 kV)	Ckm	267,044	90,000	357,000
Substation capacity	MVA	386,339	154,000	540,339
National grid	MW	28,000	27,350	55,350

Table 11.18

Capacity Addition Plan (Transmission) for 2012-2017

Sector	*2012-13*	*2013-14*	*2014-15*	*2015-16*	*2016-17*	*Total*
Transmission lines (Ckm)	18,000	20,000	20,000	16,000	16,000	90,000
Substation capacity (MVA)	26,000	35,000	33,000	32,000	28,000	154,000
Interregional transmission (MW)	5,500	5,800	5,450	5,300	5,300	27,350

Assessing the Distribution Position

Distribution sector is marred by high AT&C losses, and the uneconomical rural power lines. The AT&C losses is on the account of poor quality infrastructure, as well as due to the power theft, and poor accountability.

The Power Finance Corporation survey shows an increase in the losses of the state utilities even after accounting for the subsidy. The current revenue realised does not even cover the average cost in many cases.

Table 11.19

Distribution Networks Capacity and Targetted Additions

Infrastructure Planned	*Unit*	*Current Capacity/Length (March, 2008)*	*Eleventh Plan Targets*	*Twelfth Plan Provisional Estimates*
33 kV	Ckm	335,000	150,000	180,000
11 kV	Ckm	22,38,000	675,000	750,000
LV	Ckm	43,28,000	675,000	800,000
Step-down (33/11 kV)	MVA	639,000	130,000	116,000
Distribution transformers (11/0.4 kV)	MVA	285,000	162,000	140,000

Table 11.20

Capacity Addition Plan (Distribution) for 2012-2017

Infrastructure Planned	*Unit*	*2012-2017 Estimates*
33 kV	Ckm	160,000
11 kV	Ckm	650,000
LV	Ckm	720,000
Step-down (33/11 kV)	MVA	104,000
Distribution transformers (11/0.4 kV)	MVA	125,000

Several high profile measures are being taken to overcome the distribution losses. High voltage distribution, aerial bunched cabling, through the open access regime, anti-theft measures, IT enabled audit and monitoring improvement etc., are the most important measures that are being carried out with the support provided through the R-APDRP which was discussed in the previous section.

State level distribution companies (DISCOMs) have taken several initiatives to reduce AT&C losses and improve operational efficiency.

These initiatives include use of technologies such as aerial bunched cabling (ABC), geographical information system (GIS) and consumer indexing, high voltage distribution system (HVDS), analytics software, pre-paid metering, automatic meter reading (AMR), unmanned substations, business process re-engineering, deployment of SCADA/DMS and real-time energy auditing and accounting. Stringent theft control measures are also key to reduction of theft. As of November 2010, 34 states have set up special courts and 11 states have set up special police stations. Other best practices include spot billing, AMR, meter reading through common meter reading instrument (CMRI), outsourcing of O&M.

Gujarat and Rajasthan have taken feeder segregation and feeder in programmes respectively which has resulted in reliability of power in rural areas, and load management and reduction of losses. It is also seen that where 100 per cent metering supported by energy accounting has been completed by DISCOMs, there has been substantial reduction in losses. Privatisation of the distribution services in Delhi, and the introduction of the input-based franchising has reduced the AT&C losses from 54 to 20 per cent, has substantially reduced the transformer failure and has improved the audit and accountability.

Besides, the National Mission for Enhanced Energy Efficiency (NMEEE) is contributing to the use of modern infrastructure that enables the bolstering of performance, reduction in AT&C losses, and demand-side management measures which save power as well as avoid the addition to the generation capacity.

The below table discusses this phenomenon.

Table 11.21

Expected Savings and Avoided Capacity from NMEEE Initiatives

Sl. No.	*Initiative*	*Industry Investment (Rs Crore)*	*Fuel Saving (Mtoe)*	*GHG Emission Saving (Mn Tonnes CO_2)*	*Avoided Capacity (MW)*
1.	PAT (performance achieve transfer)	30,603	9.78	26.21	5623
2.	DSM	44,000	13.22	72.75	14335
	Total	74,603	23.00	98.96	19,958

Source: Comptroler, and Auditor General Report on the NTPC (National Thermal Power Company, or The National Power Company).

Vendor Ecosystem

The vendors in India are not able to fulfill their equipment manufacturing and commissioning targets.

BHEL is falling short of the target by 40 per cent. Against the planned commissioning of 17,338 MW in the year 2009-10, only 10,450 MW was commissioned. Similarly in this financial year, the shortfall is likely to be about 50 per cent. BHEL's stated increase in the capacity to 20,000 MW in the year 2012 is nowhere sufficient.

Hence, Government of India is promoting the joint ventures, as well as the entry of the private equipment manufacturers.

There is a need for joint ventures in the domestic manufacturing because foreign equipment tend to break down due to lack of local support expertise and due to their lack of adaptability to the varied Indian terrain. Foreign manufacturers with repair and maintenance facilities in India are allowed 5 per cent of the market share. The ultra high voltage and high voltage cables, transmission lines, switch gears and other equipment manufacturing is being ramped up in India, and 50 per cent of the purchase is to be from Indian suppliers with foreign collaboration, subject to the provision of additional warrantee.

Besides, the power procurement contracts need to at least be of one year duration, so that transparency can improve and customers can enjoy the fruits of reduced prices.

Energy Security

In India, energy security is determined by the fuel security, energy efficiency measures, environmental considerations and technological upgradation issues. In India, coal is the major source of power. Domestic coal supply falls short by about 10 per cent *viz.*, only 363 million tonnes is supplied as against the target of 401 million tonnes. There are major problems in the establishment and running of the coal mines.

Coal mines need to be established after acquiring land, providing relief, rehabilitation services, obtaining environmental clearances, etc. All these processes consume years of time in India. Power generation should also comply with environmental standards. Power transmission faces the

'Right of the Way' problem. Power distribution needs comprehensive modernisation as discussed above. Price of coal varies substantially. Captive coal mines can provide coal at Rs 500-600 per tonne. Coal India Ltd.'s mines provide coal at Rs 1,200, and imported coal costs Rs 3,000-4,000 per tonne. The estimates are based on the calorific value.

Natural gas has a small role to play in power generation, and a larger role in transport. Up to 5 per cent of India's power would be produced using the natural gas. Up to 75 per cent of natural gas is supplied by the public sector oil marketing companies (OMCs). The importance of private sector is growing rapidly with the discovery of gas reserves, and production from the Krishna Godavari Basin, and from other sources. However, gas prices are highly volatile, and are subject to the spot market volatilities to a large extent.

The energy-efficiency measures adapted in India under the leadership of the Bureau of Energy Efficiency (BEE) could reduce the capacity expansion needs by about 10,000 MW, or by about 3 per cent in India. Similarly, the low carbon growth strategies would help reduce the use of low PLF, highly polluting technologies.

There is immense scope for the adoption of the clean coal technologies *viz.*, the supercritical and the ultra supercritical technologies, so that the PLF increases up to 95 per cent and the plant shut down period gets reduced substantially. There is increased scope for increasing the PLF of the renewable energy sources and reducing the initial capital cost through R&D strategies and initiatives. Nuclear power will result in 50 per cent cost reduction on a per unit basis once the plant operates for beyond 10 years. Nuclear power plants can operate for over 60 years if sufficient maintenance is undertaken.

Manpower needs to be augmented in the supercritical and ultra supercritical areas of thermal power, high technology welding, fitting, and machinery operations, and in the field of construction and operation of high capacity and high voltage transmission and distribution infrastructure.

Stakeholders

There are multiple stakeholders whose responsibilities and goals together form the vision and the strategy of the power sector. These stakeholders collectively implement the strategy to achieve the vision of the power sector.

- The Ministry of Environment & Forests, the Ministry of Coal, the Ministry of New and Renewable Energy, Ministry of Petroleum and Natural Gas and the Department of Atomic Energy (DAE), Ministry of Finance and Ministry Power; state governments.
- Planning Commission, the Central Electricity Authority (CEA), Central Electricity Regulatory Commission (CERC) and different State Electrical Regulatory Commissions (SERCs).
- Central and state PSUs including National Thermal Power Corporation (NTPC) and National Hydroelectric Power Corporation (NHPC), Power Grid Corporation of India Limited (PGCIL), the Central Transmission Utility (CTU), State Transmission Utilities (STUs), state utilities, distribution companies.
- National Power Training Institute for capacity building and Central Power Research Institute (CPRI) for R&D.
- Power Finance Corporation (PFC), Rural Electrification Corporation Limited (RECL) and others.
- Financial institutions (including private).
- Independent Power Producers (IPPs), private players in the transmission sector and in the distribution sector.
- Bharat Heavy Electricals Limited (BHEL), other boiler-turbine-generator (BTG) suppliers, and all other suppliers other equipment suppliers (domestic and international).
- All consumers including industrial, agricultural and services sector users and the citizens of India.

Core Learning Agenda

1. There is a need to ensure supply of power and tariff adequacy through technological, market development and monitoring mechanisms.
2. Develop price discovery mechanisms: Competitive bidding process across value chain can ensure reasonable prices for end-customer further.

3. Improve financial health of state utilities, particularly DISCOMs: Rationalisation of tariffs, reducing AT&C losses and adequacy of investments are focus areas. Broad set of reforms are needed at utility level. These reforms should be carried out keeping in view the political economy of the state. Views of all stakeholders including employees of utilities must be taken into account. Interests of all stakeholders including consumers are necessary for success of distribution reform and participation of private sector.
4. Develop strong project management and execution capabilities: Project management tools to ensure timely completion of projects needs to be reinforced in both public and private sectors. Best practices should be adopted in both the sectors.
5. Ensure adequate fuel supply at both macro and micro level: securing fuel supply (both coal and gas) through optimised mix of domestic and imported fuel, long-term contracts and captive blocks allotment and mobilisation are key.
6. Develop measures for timely equipment availability particularly for Balance of Plant (BoP): Initiatives to boost augmentation of equipment manufacturing capacity in India, particularly for BoP, high voltage lines and other latest technology equipments will ensure timely domestic availability.
7. Vendor contract awarding process: There is a need for contract diversification. Care will have to be taken to avoid dependence on few suppliers for most contracts leading to vendor overload.
8. Develop learning organisation by sharing of best practices across the sector initiatives that provide a platform for experts to share best practices that can be deployed across the sector.
9. Develop specialised skills and talent: We must continue to build upon current capabilities and expertise, with special focus on specialised skill gaps identified.
10. Adopt latest technologies: As discussed earlier, starting with a small base, we have a unique opportunity for technology leapfrogging. We must evaluate and promote usage of latest technologies (e.g., USC, ultra high voltage (UHV) system, smart grids, etc.).

11. Build an environmentally conscious and sustainable growth agenda: Efficiency improvement can not only help in lowering costs and making more power available but also contribute to environmentally friendly growth agenda.

Strategy Blueprint of the Ministry of New and Renewable Energy

The Ministry of New and Renewable Energy (MNRE) plays a key role in augmenting the generation, and use of renewable energy in India. At present, over 11 per cent of the total electricity consumed eminates from renewable energy sources. The strategy blueprint aims at taking this figure to 20 per cent by the year 2020, and over 50 per cent by the year 2032, including the nuclear power. The ministry identified the several key requirements to realise development of renewable energy and making it available at low price.

Manpower development, HRD, training, provision of institutional finance, promotion of cogeneration on BOLT basis, provision of generation-based incentives (GBI) are some of the important initiatives. They were discussed in Chapter 9. However, we need to understand the important measures taken by the ministry towards the evolution of effective strategy for the renewable energy sector.

Reform Measures and Policy Initiatives Taken during 2010-11

Towards the effective implementation of the Jawaharlal Nehru National Solar Mission (JNNSM), the Government took the following initiatives.

Guidelines for technical qualification and accreditation of system integrators were issued. *r a* r ◇⋞ for refinancing of off-grid solar application through NABARD were issued. Guidelines for the implementation of GBI through IREDA were issued. Testing and certification of turbines was made mandatory and its scheme was expanded to include direct proposals from manufacturers and suppliers apart from the state nodal agencies.

To encourage development of bagasse cogeneration projects in cooperative sector sugar mills, a scheme for supporting the initiative and implementing on BOOT/BOLT basis by IPP/joint venture companies was introduced.

Plan for Renewable Energy for Ladakh Region

A project entitled 'Ladakh Renewable Energy Initiative' to minimise dependence on diesel in the Ladakh region and to meet power requirement through local renewable sources was commenced. The project aims at meeting power requirements through micro hydel and solar photovoltaic power projects/systems and use of solar thermal systems for water heating/heating/cooking requirements. The project will be implemented in three and a half years with a total cost of Rs 473 crore. It includes the setting up of 30 small/mini hydel projects with 23.8 MW of aggregate capacity at a total cost of Rs 266 crore.

Innovative Projects

Several projects aimed at fostering innovations in the renewable energy sector are initiated.

Some of the initiatives include institution of renewable energy chairs in higher educational institutions, laboratory and library facilities of educational institutions for imparting quality education, developing and incorporating model renewable energy courses in the course curricula of seven trades of two-year ITI course, increasing the scope of renewable energy fellowship scheme, institution of national solar science fellowship programme. In addition to innovation in the area of renewable energy, Ministry has initiated process of developing Centre for Innovation, Incubation & Entrepreneurship (CIIE) of IIM Ahmedabad (IIMA) as Centre of Excellence for scouting innovation, incubation and enterpreneurship development in renewable energy sector.

Pitfalls in Project Management: The Audit Approach

The Comptroller and Auditor General (CAG) of India is responsible for auditing and holding the Government of India, its associated ministries and departments, and enterprises, along with those of the state governments.

The National Thermal Power Corporation, or the NTPC is the premier power generation company in India. By generating over 33,000 MW of power, it is responsible for 18 per cent of the total power generation in India. The NTPC is aiming at commissioning projects aimed at generating a further 42,000 MW by the year 2017. By the year 2012 itself, NTPC is supposed to commission projects to generate 17,000 MW taking the capacity to 50,000 MW.

NTPC is facing lots of problems in the project execution and commissioning. Some of the problems are internal to the NTPC, and some of them are external.

The CAG has provided a critical review of the problems and has cited several avoidable problems.

Company Profile

NTPC Limited (Company) was incorporated in November 1975 with the objective to plan and promote development of thermal power in the country. In May 1998, NTPC diversified its objectives to include new business activities like development of hydropower and power through non-conventional/renewable energy sources. As on 31 March 2010, NTPC with an installed power generating capacity of 28,902 MW from 15 coal-based thermal power stations and seven gas/liquid fuel-based power stations was the largest thermal power generating company of the country. NTPC contributed 28 per cent of the total electricity generation of the country as of 31 March 2010. NTPC is a government company wherein Government of India holds 84.50 per cent of the total equity.

Capacity Addition Programme of NTPC

In line with the objective of National Electricity Policy (February 2005) to increase annual per capita consumption of electricity from the existing level of 631 units to 1,000 units, NTPC planned (April 2007) a Capacity Addition Programme to become a 50,000 MW Company by 2012. Since, NTPC had installed capacity of 27,404 MW upto 31st March 2007, the company planned to increase the capacity by 22,600 MW (83%) during next five years (2007-2012). However, due to non-fructification, rescheduling or substitution, the company revised (2007-08) its target downward to 22,430 MW to be achieved during 2007-2012. In order to achieve this target, NTPC decided in (2006-07) to adopt a multipronged growth strategy.

The CAG stated in November 2010 that the capacity addition target of 22,430 MW was approved neither by the Planning Commission nor by the Ministry of Power. Analysis should be done only against the target for capacity addition of 17,760 MW fixed by Planning Commission for NTPC for Eleventh Plan.

The CAG does not agree with the management as the capacity addition target of 22,430 MW was set by the company itself as per the corporate plan. The company had in fact identified projects to augment capacity for 22,430 MW during 2007-2012 and had also initiated action on these projects.

Progress of Capacity Addition Programme

Of the target of 22,430 MW fixed involving 24 projects, the company commissioned only five thermal projects involving 4,220 MW (19%) till 31/03/2010.

Thus, against the target of 22,430 MW, company would be able to achieve only 9,220 MW resulting in a shortfall of 13,210 MW (59%). Analysis of progress made so far reveals that although the company anticipates to achieve additional 4,530 MW involving six projects on best effort basis, commissioning of these projects appears to be difficult by March 2012 as present status of progress of work is far from satisfactory. Further, the company could achieve only 500 MW against 1,500 MW planned on best effort basis during previous capacity addition programme (2002-2007).

Master Network Aids the Timely Completion of the Project

Master network in respect of every project had to be developed indicating matching programme or each balance of plant (BoP) contracts such that synchronisation of the project is achieved as per schedule agreed for main plant contract. A review of awarding of 103 BoP contracts revealed that in 74 contracts, master network dates were not adhered to and delays ranging between one and 24 months were noticed as is evident from the Table 11.22.

Table 11.22

Range of Delays in Project Completion

Range of Delay (in Months)	*Number of Contracts*	*Name of the Projects*
1 to 3 months	27	Barh-II, Farakka-III, Kahalgaon-II, Kobra-III, Dadri-II, Simhadri-I, Sipat-I, Sipat-II, Jhajjar
More than 3 up to 6 months	23	Barh-I, Barh-II, Bangalore, Farakka-III, Kahalgaon-III, Korba-III, Dadri-II, Simhadri-II, Vallur
More than 6 but up to 9 months	7	Barh-I, Kahalgaon-II, Dadri-II, Simhadri-II, Vallur
More than 9 up to 12 months	10	Barh-I, Bongaigaon, Farakka-III, Korba-III, Nabinagar, Vallur
More than 12 up to 24 months	7	Korba-III, Nabinagar, Vallur

Source: Same as Table 11.21.

Table 11.23

Thermal Projects

Scheduled COD	*Actual COD*	*Delays in Completion (in Months)*	*Financial Progress/ Status as on March, 2010*
July, 2007	December, 2008	17	Completed
June, 2007	March, 2010	33	Completed

Sipat-II (2 x 500 MW)

Unit No.	*Scheduled COD*	*Actual COD*	*Delays in Completion (in Months)*	*Financial Progress/Status as on March, 2010*
4	September, 2007	June, 2008	9	Completed
5	March, 2007	January, 2009	9	

Note: COD: Commercial operation rate.

Source: Same as Table 11.21.

NTPC's Hydro Projects

The date of commissioning of the hydro projects was to be worked out from the date of award of the first contract of main civil works package as per techno economic clearance (TEC) issued by CEA.

Table 11.24

Delays in the Commissioning of Hydro Projects

Sl.No.	Name of the Project	Date of TEC	Date of Award of First Package of Main Civil Work	Time Taken from TEC to Award
1.	Koldam	30-6-2002	12-12-2003	18
2.	Loharinag-Pala	11-8-2004	6-7-2006	22
3.	Tapovan-Vishnugad	11-8-2004	28-11-2006	27

Source: Same as Table 11.21.

Capacity Addition Programmes in the Eleventh Plan: Between 2007 and 2012

Table 11.25 discusses the approved projects across the states by source.

Table 11.25

Details of the Projects Identified by NTPC for Capacity Addition Programme (2007-2012)

Sl. No.	Name of Project	State	Capacity Envisaged (MW)					Status Project (Greenfield or Expansion)
			Hydro	Coal	Gas	Wind	Grand Total	
Project to be executed by NTPC exclusively:								
1.	Koldam	Himachal Pradesh	800	-	-	-	800	Greenfield
2.	Lahorinag-Pala	Uttarakhand	600	-	-	-	600	Greenfield
3.	Tapovan-Vishnugad	Uttarakhand	520	-	-	-	520	Greenfield
4.	Dadri-II	Uttar Pradesh	-	980	-	-	980	Expansion
5.	Sipat-I	Chhattisgarh	-	1980	-	-	1980	Expansion
6.	Sipat-II	Chhattisgarh	-	1000	-	-	1000	Greenfield
7.	Korba-III	Chhattisgarh	-	500	-	-	500	Expansion
8.	Mauda	Maharashtra	-	1000	-	-	1000	Greenfield
9	Simhadri-II	Andhra Pradesh	-	1000	-	-	1000	Expansion
10.	Kahalgaon-II	Bihar	-	1000	-	-	1000	Expansion
11.	Barh-I	Bihar	-	1980	-	-	1980	Greenfield
12.	Farakka-III	West Bengal	-	500	-	-	500	Expansion
13.	Barh-II	Bihar	-	1320	-	-	1320	Expansion
14.	Bongaigaon	Assam	-	750	-	-	750	Greenfield

contd...

...contd...

Sl. No.	*Name of Project*	*State*	*Capacity Envisaged (MW)*					*Status Project (Greenfield or Expansion)*
			Hydro	*Coal*	*Gas*	*Wind*	*Grand Total*	
15.	Rihand-III	Uttar Pradesh	-	1000	-	-	1000	Expansion
16.	Vindhyachal-IV	Madhya Pradesh	-	1000	-	-	1000	Expansion
17.	North Karanpura	Jharkhand	-	660	0	0	660	Greenfield
18.	Guledagudda	Karnataka	-		-	100	100	Greenfield
	Total		1920	14670		100	16690	
Projects to be Executed by NTPC through its Joint Ventures/Subsidiaries								
19.	Bhilai (JV with SAIL)	Orissa	-	500	-	-	500	Expansion
20.	Jhajjar (JV with HPGCL)	Haryana	-	1500	-	-	1500	Greenfield
21.	Ratnagiri GPPL (JV with GAIL, banks and FIIs)	Maharashtra	-	-	740	-	740	Revival Project
22.	Vallur (JV with TNEB)	Tamil Nadu	-	1500	-	-	1500	Greenfield
23.	Nabinagar (JV with Indian Railways)	Bihar	-	1000	-	-	1000	Greenfield
24.	Muzaffarpur Expn.	Bihar		500	-	-	500	Expansion
	Subtotal	-		5000	740	-	5740	
	Grand Total		1920	19670	740	100	22430	

Hydro Projects

The date of commissioning of the hydro projects was to be worked out from the date of award of the first contract of main civil works package as per TEC issued by CEA. In order to achieve the capacity addition programme (2007-2012), the first contract for main civil work of hydro projects should have been awarded after TEC at the earliest. However, it was observed that the company had taken inordinate time ranging from 18-27 months for awarding of fist package of main civil works from the date of TEC as is evident from Table 11.26.

Table 11.26

Hydro Projects

Sl.No.	*Name of the Project*	*Date of TEC*	*Date of Award of First Package of Main Civil Work*	*Time Taken from TEC to Award*
1.	Koldam	30-6-2002	12-12-2003	18
2.	Loharinag-Pala	11-8-2004	06-07-2006	22
3.	Tapovan-Vishnugad	11-8-2004	28-11-2006	27

NTPC does not provide the breakup of the tasks, operations, and the costs for the capacity addition projects.

The direct result of the colossal cost overruns is manifested in the downgrading of the Muzaffarpur Thermal Power Project from 1,500 MW to 390 MW *viz.*, unit sizes of 750 MW each to 195 MW each.

Audit Framework

Scope of Audit

This performance audit covers all the activities from conceptualisation to implementation of all the projects selected by NTPC and its joint ventures/subsidiaries for adding a capacity of 22,430 MW between April 2007 and March 2012.

Audit Objectives

In order to analyse the reasons for slow progress of work and to assess the effective and economic implementation of the capacity addition programme, a performance audit covering activities from conceptualisation to execution of projects by NTPC and its joint ventures (JVs)/subsidiaries to achieve the targets of capacity addition set for 2007-2012 was undertaken. Audit objectives were to assess whether:

- The projects were selected and formulated considering all relevant factors;
- Projects and contracts were managed with due economy and efficiency;
- There was an effective monitoring mechanism at all levels and
- The delay in completion of projects involved any financial impact.

Thermal Projects

For achieving the Capacity Addition Programme by March 2012, the company should have ensured during the initial period that there was no slippage in awarding of the main plant packages. It was observed that due to excessive time taken during pre-order activities, awarding of main plant contracts in six projects was delayed. As a result, capacity of 5,210 MW right away slipped beyond March 2012 by 7 to 20 months as is evident from the table given below:

Table 11.27

Delay in the Completion of Thermal Projects and the Reasons for Delay

Sl. No.	*Name of the Project*	*Capacity to be Added (in MW)*	*Date of Award of Main Plant Contract*	*Scheduled Date of Completion as per Investment Approval (Unit-I/Unit)*	*Delays Beyond March 2012 in Scheduled Completion (in Months)*	*Reasons of Delay in Award of Main Plant Contract*
1	Barh-II (2 x 660)	1320	14-10-2008	January 2013/ November 2013	20	Delay in finalisation on main plant contract
2	Mauda (2 x 500)	1000	28-11-2008	May 2012/ November 2012	8	Delay in acquisition of land
3	Rihand-III (2 x 500)	1000	18-2-2009	July 2012/ January 2013	10	Delay in initiation of pre-order activities
4	Vindhyachal-IV (2 x 500)	1000	18-2-2009	July 2012/ January 2013	10	Delay in initiation of pre-order activities
5	Vallur-II (1 x 500)	500	28-7-2009	January 2013	10	Delayed decision about project size
6	Muzaffarpur	390	12-3-2010	October 2012/ January 2013	10	Delayed decision about project size
	Total	5210				

Table 11.28

Statement Showing Details of Critical Packages (in Addition to Sample) Awarded with Significant Delay

Sl. No.	Name of Project	Name of Project not Awarded till December, 2009	Scheduled Date of Award	Date of Award	Delay in Months	Scheduled Completion Period
1	Barh-II	Coal Handling Plant	30-10-2008	25-1-2010	15	39
		Circulating Water & Make up Water Civil Works	14-10-2008	26-2-2010	16	30
		Chimney	14-11-2008	5-11-2009	12	30
		Station Control & Instrumentation	14-12-2008	8-10-2009	10	41
2	Bongaigaon	Coal Handling Plant	28-10-2008	20-1-2010	15	30
		Ash Dyke	17-10-2008	Not awarded	19	-
3	Mauda	Coal Handling Plant	15-5-2009	22-1-2010	8	36
		Control & Instrumentation	6-7-2009	Not awarded	11	-
4	Rihand-III	Coal Handling Plantg	30-5-2009	27-1-2010	8	37
		Ash Dyke	March, 2008	11-11-2009	19	-
5	Vindhyachal – IV	Coal Handling Plant	24-6-2009	Not awarded	11	-
6	IGSTPP, Jhajjar	Ash Dyke	March, 2008	11-11-2009	19	-
7	Vallur STTP	External Coal Handling Plant	12-12-2007	30-3-2010	27	20
		Internal Coal Handling Plant	15-6-2008	29-3-2010	21	32
		Ash Dyke	15-7-2008	11-2-2010	18	-

The total cost of 21 approved projects between 2002 to 2010 involving 21320 MW was 1,09,319.19 crore, or about Rs 5.2 crore per MW. But a cumulative expenditure of Rs 50,853.19 crore has already been incurred up to March 2010. This demonstrates the 'financial profligacy' of the NTPC project execution team. The project cost shot up to about Rs 12.5 crore per

MW, which is a colossal waste of resources. This cannot elevate NTPC up to best in the World.

Audit Criteria

The performance of NTPC was assessed against the following criteria:

- National Electricity Policy,
- Corporate Plan of NTPC,
- Minutes of meetings of the Board of Directors (BOD) and sub-committees,
- Feasibility Reports (FR)/Detailed Project Reports (DPRs),
- Work and Procurement Policies and Procedures,
- Contract Agreements and
- Best practices adopted by the industry.

Audit Meeting

The Audit held a meeting with the NTPC Management on 7 July 2010 wherein, the scope, objectives, criteria of audit and audit sample were discussed. Audit examined relevant records based on which preliminary observations were issued to the management and the replies of the management, wherever received, were considered while drawing audit conclusions which have been narrated in subsequent Chapters. After receipt of responses from the Management, an exit conference was held on 18th November 2010 and further clarifications and comments made by the Management during this interaction were also considered while finalising this report.

Audit Sample

The projects covered under Capacity Addition Programme of NTPC are to be completed by March 2012 and, therefore, some of the projects were still under award or execution.

As such a representative sample of the contracts was drawn from list of contracts awarded for executing the projects using Interactive Data Extraction and Analysis (IDEA) software for examination of contract awarding activities. Out of 442 contracts valuing Rs 58,752 crore awarded between April 2003 and March 2009 for Capacity Addition Programme, 150 contracts (34%) valuing Rs 50,706 crore (86%) were selected for review of award activities. As regards execution, all 24 projects under execution were selected for examination.

Key Findings

The CAG flagged the NTPC management mainly on the grounds of site finalisation, technology selection, cost overruns, and lack of results in the monitoring of the projects, besides on methodological grounds.

About 6,130 MW of previous capacity addition plan was shifted to the Eleventh Plan. The Management's reply was that the feasibility study forms part of the project management is not accepted by CAG.

Serious delays in site acquisition as a result of disputes on the coal availability with ministry of coal, JV with Ministry of Railways, resettlement and rehabilitation (R&R) issues etc., cause a delay of anywhere between 48 months and 125 months before the project execution itself.

NTPC failed to adopt supercritical technologies in over 80 per cent of the projects. Subcritical technologies are costly, occupy at least 25 per cent more land and are environmentally unfriendly, besides slower in execution.

Besides, in the case of Farakka project, NTPC did not claim about Rs 1,500 crore of deemed export benefits.

The monitoring of the projects is taking place at the manager's regional director's, the director's, and also at the managing director's level. Yet, there does not seem to be any time saving. Only the Jhajjar project is executed on the 'best practice principle.'

The cost overrun in the 660 MW plant of NTPC seems to be over 20 per cent compared to the private sector projects.

The above discussion suggests that NTPC is least likely to achieve the target of 75,000 MW of power production by 2017 and is also least likely to achieve the target of 60,000 MW.

Some Criticisms over the CAG's Approach

CAG follows the approach of auditing the corporation. This is least likely to hold the project managers accountable, let alone the highest management cadre. CAG discusses about the financial overruns, while at the same time frees the vendors and contractors from blame. On one hand, the CAG blames the management of delays and inefficiencies. But the possible corruption and unlawful dealings angle is not investigated.

Concluding Remarks

A comprehensive strategy is required to make India's power sector qualitative, reliable, and provide adequate and cost-effective services to the consumers. India lags behind in the technical, technological and financial spheres. Besides the project execution capability of India's public sector enterprises is far short of those in the private sector.

A comprehensive study needs to be conducted on the reasons behind the poor performance of the public sector management in this respect, identify key remedial measures, and implement them. Salary component could not be the reason. The salaries paid by the private sector to the key personnel is more than in public sector.

Perhaps, the reasons have to do with incentive structures and motivational factors.

12 Conclusions and Recommendations

The ideas and the understanding derived from this book are valuable from the viewpoints of policymaking, career planning and for developing a futuristic understanding. The chapter summarises the conclusions and presents useful recommendations to the Government and to the readers. The world is facing various energy risks. They are understood as the systemic risks and the market risks.

The world has to move away from the oil, gas and coal-based to nuclear and renewable-based energy use. Hydropower potential is fast approaching stagnation. The Herfindahl Index shows the extent of source diversification. Major economies of the world face a threat to energy security.

USA, Japan, the European Union, all depend on imported energy to a greater extent. They all face the price, geopolitical and scarcity risks pertaining to the fossil fuel-based energy resources. Hence, all these countries opt to diversify into the renewable energy sphere, including the solar, the wind, the fuel cell technologies etc. The recent nuclear disaster in Japan results in a slowdown in the dependence on the nuclear energy. Heavier reliance, thus will be emphasised on solar energy, as the wind power potential is limited.

India's energy independence pertaining to the fossil fuels is a distant phenomenon which possibly would not be realised within this century. India's strategic disadvantage arising out of geopolitical situation in sourcing for oil has to be overcome by establishing stable and durable economic relationship with reliable partners like the Russia, the Central Asian countries etc. Energy efficiency has to be increased and technological advancements in the arena of energy exploration, development and production along with consumer appliances has to be adapted on a grander scale to improve the energy security and to continue the process of reduction in energy intensity.

Indo-Russian defence ties and rapidly growing bilateral trade has to be strengthened through better energy cooperation. There is a possibility to strategise China as a buffer between India and Russia to enhance secured energy imports from Russia to India and foster better business and economic cooperation. China is the biggest trading partner of India with bilateral trade touching US $25 billion. Whereas, Indo-Russian bilateral trade was only worth $10 billion by the year 2010. By securing the trade routes between India and Russia both overland and through the South China Sea, both India's as well as China's economic security can be ensured on a sustainable basis. To fulfill this objective, Indo-Russian-Chinese naval cooperation has to be fostered through mutual trust and goodwill.

Fossil fuel and bioenergy-based consumption of energy items coexist in India. Hence, India has to go a long way to be fully recognised as a modern country. Still, 70 per cent of people cook using the firewood and chips and over 40 per cent of the rural households lack access to the electricity. While the modernisation efforts have to proceed unimpeded, the initiatives should be based on cleaner technologies like the renewable energy technologies so that the pollution can be controlled and sustainable development can be ensured. At the outset, the use of clean coal technologies become important.

India is embarking upon major plans for modernisation and expansion of the oil refining capacity. This will help in the generation of valuable foreign exchange through the export of the refined products, and help circumvent the high cost of crude purchase.

However, the refining capacity is slated to grow in the established centres like Singapore as well. Hence, India has to evolve into a comprehensive trading, financial and commercial hub for oil that serves the Indian Ocean Rim region. Mumbai is well placed to assume the role.

India has to ensure the penetration of electricity and electrical appliances while at the same time reducing deforestation and the use of kerosene. The transmission and distribution losses of the electricity should be controlled to a large extent, and people and functionaries in the high loss states should be sensitised. Peak time management of electricity should be ensured using various physical and financial instruments of contract.

The world has made great strides in enhancing energy efficiency. In the past 100 years, the world has saved 60-75 per cent of gasoline in running a motor car. Similarly, the energy intensity of the economies has reduced by more than 2/3rd. Even in India, the energy efficiency has reduced tremendously. There is a great scope for reducing wasteful energy consumption and promoting the energy efficiency. Even the municipal solid waste can be converted to energy through the biomass power plants.

R&D initiatives in India have made some strides in promoting energy efficiency in domestic, commercial and industrial establishments. In the household, modern construction methods can enhance the energy efficiency by up to 40 per cent through better insulation and through the prevention of the leakage. In the commercial and in the industrial establishments, up to 30 per cent and up to 20 per cent efficiency can be achieved.

The usage of public transport and particularly the rail transport is heavily recommended because the per passenger fuel saving is up to 40 per cent. Moreover, the direct use of coal and the diesel wastes over 60 per cent and over 50 per cent of energy respectively. Hence, it is recommended that all the transportation should be run using the electricity. Even the industrial establishments should use oil to the best possible extent.

However, oil has a role to play for another 40-60 years and natural gas for 60-100 years respectively in the area of transportation and industrial use depending on the estimates. The pressing concerns faced by the world are the sustained increase in the production and supply of oil and natural gas.

Technological innovations are and should take place at the wellhead to make the hitherto uneconomical oil resources more economically extractible. The supply lines should be secured from geopolitical risk of attacks and vandalisation. Suppliers should have the economic incentive to supply oil and natural gas. Therefore, long-term price contracts should encompass all the possible costs and risks. At the same time, the pricing of oil and natural gas should not be prohibitory of economic activity.

In the case of India, a sustained 8-9 per cent growth rate in the economy should be maintained until 2032 and 6-7 per cent until 2050-2060 to overcome poverty and educational gap. Strategic trade ties should be established by India in the East Coast and in the West Coast to tap the natural gas resources of Indonesia, Malaysia and Qatar at economically competitive price. India is attempting to overcome the foreign exchange cost of oil and natural gas import by trying to open up the refineries for export processing thus contributing the much needed foreign exchange.

To establish geopolitical security of oil and natural gas, the US/Middle Eastern ties and the Russian/European and the Indo-Russian ties are being strengthened. A limited amount of taxation is being enforced to collect funds for renewable energy research that smoothens the transition to renewable energy on an economic basis.

The transition to renewable energy would take a long time viz., another 20 to 40 years. In the mean time, scarce resources should be utilised optimally. As the trading and trading instruments pertaining to the oil and the natural gas have developed, trading in electricity and also the appropriate mechanism and instruments *viz.*, the models and contracts have developed. Power trading has come to stay. Various instruments of risk minimisation are being used to smoothen the flow of power and to hedge against the financial losses otherwise to be suffered by the utilities.

Power trading mechanisms such as the financial and physical instruments used in 'over the counter' and exchange-based trading are improving over time and the demand and supply are being tracked on a daily and even on an hourly basis.

The electronic, communication and computer-based tracking and monitoring systems are enabling the timely detection of the problem, its repair and maintenance, ensuring proper and timely payment of the bills and almost instant communication between the power generation companies and its customers—including the large end-use customers.

In the case of India, power trading has taken off a decade ago. Over 80 per cent of trading takes place through the long-term contracts. The volume of exchange-based power trading is miniscule in India. However, there is a potential to enhance the exchange-based trading to manage the peak time power and to conduct 'load shedding'. Exchange-based contracts are so far short-term in nature and are at the most 'week-ahead'. Only about 350 large customers are involved in the exchange-based trading.

The need for abundant and sustainable energy resources was recognised right at the time of India's independence by the scientists and policymakers. As a first step, the nuclear programme was envisaged to be developed. India's peaceful nuclear programme has evolved through various difficulties like technology denial and the problems involved in 'reinventing the wheel.' Even sanctions against the import of technologies and spare parts were in force for over three decades as enforced by the USA.

But there is a transformation in the US' mindset that hitherto considered nuclear cooperation with India is a risk in terms of arms race and nuclear proliferation. The present state of understanding reflects the 'win-win' situation through Indo-US cooperation. This is especially because Indian scientists are confident in the ability of Stage-III thorium-based fast breeder reactors fully meeting the energy needs of the world although after many decades from now.

The increased self-reliance and self-confidence of India's nuclear establishment enabled the US to understand the 'win-win' situation arising out of the Indo-US nuclear deal. Commercial principles should override the foreign policy agenda of US and through joint enterprise. US and India should adopt and utilise the best technologies and practices of all the companies producing nuclear reactors.

Nuclear energy is not a sufficient form of renewable energy at least for the next many decades. At the most, India can increase the share of nuclear electricity from 3 per cent to 10 per cent in the next 20 years. But other renewable energy technologies such as the wind, solar and bioenergy are necessary and important in the immediate sense.

The exhaustion of the fossil fuels and the GHG emissions from the fossil fuels warrant the development and use of the renewable energy technologies. India's NAPCC necessitates the reduction of GHGs and the promotion of the CDM projects. For this purpose, various programmes aimed at promoting eco-friendly and sustainable development is envisaged in the NAPCC. Agriculture, renewable energy, green buildings, waste to energy, sustainable habitat etc., are the important components of the NAPCC. Important CDM projects concern with the cleaner use of coal and natural gas and the hybridisation of the technology. The PFBC and the IGCC are the important forms of CDM projects concerning the coal and natural gas. CDM projects can enhance the efficiency of the power plants by about 50 per cent. The National Solar Mission is part of the CDM mechanism.

Patents are the indicators of the depth and width of the technology development and absorption. The USA, Japan, Germany, UK, France and Canada lead the way in the development of original patents as well as the derived patents under the PCT regime. China and South Korea predominate in the PCT-based patents. Most countries mentioned above focus on the hydrogen and fuel cell research. USA, Germany and Japan concentrate more on the solar power. Also Germany and USA concentrate more on the wind power. China has special focus on developing the hydropower technologies. The flow of human resources can be understood by referring to the patent citations. Most Western European countries balance the brain drain and brain gain. USA is susceptible to 'brain drain' while Japan is not inclined towards any of these tendencies.

Renewable energy, especially the wind energy has made good strides in India. India is able to meet the renewable purchase obligation (RPO) of 10 per cent through the renewable energy mostly through the wind energy. The future potential of the wind, biomass and small hydropower is 4, 10 and 10 times the present level respectively. The future potential of the solar energy in India is almost infinite and has the potential to meet the energy requirements of India. Although the solar water heaters (SWHs) are growing fast in usage, the growth has to be manifold faster to be comparable with China.

Solar PV applications and solar thermal applications are being given greater impetus in India. Unlike the present domination of the off-grid applications, the paramount is given to grid-based applications by the JNNSM. Solar collector capacity is growing 'rapidly in the world and in leaps and bounds' in India *viz.*, at an annual rate of almost 25 per cent. However, India is constrained by the low temperature, low efficiency solar PV modules.

Hence, JNNSM gives impetus to R&D and manufacturing of high and medium temperature applications through refinance, tax incentives, RPOs and through incubation support. Provision of Intellectual Property Rights through international cooperation and collaboration and the gradual setting up of 'open source' knowledge bank pertaining to solar power are the highlights.

Since India's solar PV sector is dominated by 15 companies producing only up to 700 MW of modules, impetus is given for small and medium enterprises to produce and market taking the total capacity up to 8 to 10 GW by the year 2022 *viz.*, by the end of the third phase of JNNSM.

From the present 10 KW limit of SPV power plant, as envisaged by MNRE, SRC is envisaging R&D projects to set up plants up to the capacity of 100 MW and hybrid plants *viz.*, solar/wind and solar/coal of up to 150 MW respectively. For this purpose, research and technological base is necessary. Premier educational institutions *viz.*, the IITs and the IIMs are required to take the lead in the technological and managerial education pertaining to solar technology and management.

Up to 100 world class scientists and technologists, up to 1,000 good scientists and up to 1,00,000 technicians are expected to work in the solar energy sector in India by the year 2020. Appropriate monitoring and implementation mechanisms are already set in place by the JNNSM under the aegis of the Mission Steering Group, the Mission Executive Committee, the Mission Directorate and the Industrial Advisory Committee respectively.

Wind power has become the dominant source of renewable energy. However, it will be replaced by solar power as the potential of wind power is limited. Therefore, it is essential to note that over 5 per cent of the world's electricity requirements are met by the wind power. In India, the figure has already reached almost 8 per cent and the potential would be 48,000 MWh or 6 per cent of India's energy requirement in 2030.

Recommendations

The lifetime of the fossil fuels should be sustained at least until the end of their estimated life period of 40 years for oil, 60 years for natural gas and up to 200 years for coal. The development and production of unconventional oils like the tar sands and the heavy oils from the oil shales should not cause environmental problems and earthquakes.

Power trading must be transparent and the instruments of the contracts should fulfill the expectations of all the market participants. The Integrated Energy Policy must be augmented to reserve certain portion of electricity to be traded in the power exchange so that the energy wasting industries would be disincentivised.

India's thorium-based nuclear programme should be the Government priority and the civilian nuclear programme should be opened up for the private sector with due consideration for human security so that scarce government financial resources can be put to better use.

Renewable energy should be adapted at the domestic, commercial and at the industrial level in India using the most proven technology. India's scientific and technological brains must be properly utilised to foster R&D in the renewable energy arena. Even the international collaboration efforts and the education of Indian scientists and technical personnel in foreign universities and research institutions must result in gainful patent recognition for India.

The proven technologies and those technologies becoming cost-effective must be dropped from the Government incentive basket so that it's resources can be used to promote higher level of technologies in research and use. This is already recognised by the JNNSM which discusses the focus shift from the second phase to incorporate medium and high temperature applications.

Real time monitoring of the availability of new technologies should be carried out on a sustainable basis by a 'highly resourceful agency' under the public-private partnership (PPP). Efforts to integrate those technologies in the national policy and plans should be done on a continuous basis.

A comprehensive strategy is required to make India's power sector qualitative, reliable and provide adequate and cost-effective services to the consumers. India lags behind in the technical, technological and financial spheres. Besides the project execution capability of India's public sector enterprises is far short of those in the private sector.

A comprehensive study needs to be conducted on the reasons behind the poor performance of the public sector management in this respect, identify key remedial measures, and implement them. Salary component could not be the reason. The salaries paid by the private sector to the key personnel is more than in public sector. Perhaps, the reasons have to do with incentive structures and motivational factors.

Bibliography

Amineha, Parvezi Mehdi and Henk Howelingb (2007). *Global Energy Security and its Geopolitical Impediments - The Case of the Caspian Region*. PGDT 6, pp. 365-388, *www.brill.nl/pgdt*

Anantaram, Rajeev (2010). *The Development of the Power Sector in India: Issues and Prospects*. (ARI), Real Instituto Elcano, January. *Http://www.realinstitutoelcano.org/wps/portal/rielcano_eng/Content?WCM_GLOBAL_CONTEXT=/elcano/elcano_in/zonas_in/ari6-2010*

Bagla, Pallava (2007). "Implementation of Hyde Act would mean shifting of 'Goal Posts'", *The Hindu Daily, Http://www.hindu.com/2007/01/17/stories/2007011702641100.htm*

Blagov, Sergei (2007). "Russia Struggles to Sustain Partnership with India", *Eurasia Daily Monitor* 4(213) November 15. *Http://www.jamestown.org/single/ ?no_cache=1&tx_ttnews%5Btt_news%5D=33168*

Brown, Lester R. (2001). *Turning Sunlight into Electricity, Eco-Economy: Building an Economy for the Earth*. Earth Policy Institute, *http://www.earth-policy.org/Books/Eco/EEch5_ss4.htm*

Carbon Dioxide R&D Center (CDRS) (2008). *Global Warming, Its Cause and Solution*. *Http://www.cdrs.re.kr/down/cdrs_hongbo.pdf*

Charles, Ferguson D. (2008). *Reshaping the US-Indian Nuclear Deal to Lessen the Nonproliferation Losses*. Arms Control Association. *Http://www.armscontrol.org/act/2008_04/Ferguson.*

Chaudhuri, Saikat (2007). *Business Implications of the Indo-US Nuclear Deal. Knowledge@Whartonhttp://knowledge.wharton.upenn.edu/india/article.cfm?articleid=4217*

Clarke, Kieran and D. Graczyk (2010). "India's Downstream Petroleum Sector Refined Product Pricing and Refinery Investment", *Working Paper*, International Energy Agency (Australian Department of Resources, Energy and Tourism).

Dadhich, Pradeep (1999). *Indian Power Sector*. OPED: The Energy Research Institute. *Http://www.teriin.org/opet/articles/art2.htm*

Deffeyes, K.S. (2005). Beyond Oil: The View from Hubbert's Peak. Hill and Wang, a division of Farrar, Straus and Giroux, New York, pp. 1-51; 169-188.

Dwivedi, Prasoom and Parag Diwan (ed.) (2008). "Energy Conservation", *Energycopia* 3, The Energy Research Institute, Chapters. 1-4

Encyber (2007). *Geothermal Power Generation*, Encyclopedia website, *http://news.encyber.com/search_w/ctdetail.php?gs=ws&gd=&cd=&q=&p= &masterno=142782&contentno=142782*

Energy in India, *http://www.cslforum.org/India.htm*

Energy Information Administration (US) (2007). India Summary. January.

Esther, Pan and Jayshree Bajoria (2008). *The US India Nuclear Deal*. United States Council on Foreign Relations. *Http://www.cfr.org/publication/9663/.*

Global Wind Energy Council (GWEC) (2007). *Global Wind 2007 Report. Http://www.gwec.net/fileadmin/documents/test2/* gwec-08-update_FINAL.pdf

Goel, Sharad and Parag Diwan (ed.) (2008). "Energy Trading", *Energycopia* 4, The Energy Research Institute, pp. 6-33, 52-89, 121-237.

Goldemberg, Jose, Thomas B. Johansson, Amulya K.N. Reddy and Robert H. Williams (1985). *Energy for a Sustainable World*. The Wiley Eastern Limited, Chapters 1, 2, 3, 4 and 5, August.

Goswami, B.N., V. Venugopal, D. Sengupta, M.S. Madhusoodanam and Prince K. Xavier (2006). "Increasing Trend of Extreme Rain Events Over India in a Warming Environment", *Science* 314(1442): 15.

Government of India (nd). *National Action Plan on Climate Change*. Government of India Prime Minister's Council on Climate Change. Supported by the Office of the Principal Scientific Adviser to the Government of India.

———. (2009). *Position Paper on the Power Sector in India*. Department of Economic Affairs, Ministry of Finance, Government of India, December.

———. (2010a). *Annual Report*. Ministry of Power, Government of India.

———.(2010b). Tables on Renewable Energy, Wind Energy, Solar Energy, Small Hydro Power, Ministry of New and Renewable Energy, G Government of India.

———.(2010c). *Report on the Loss of Value in the National Power Company*. Comptroler and Auditor General, Government of India.

———.(2011a). *Strategy Blueprint*. Ministry of Power, Government of India.

———.(2011b). *Strategy Blueprint*. Ministry of New and Renewable Energy, Government of India.

House on Evolution of India's Nuclear Policy (1998). *Evolution of India's Nuclear Policy.* Embassy of India. *Http://www.indianembassy.org/pic/nuclearpolicy.htm*

ICRA Ltd (2006). *State Power Sector Reforms Performance*. Final Report to the Ministry of Power, Government of India, June.

IEA-SHC (2007). *Task33–Solar Heat for Industrial Processes. http://www.iea-shc.org/task33/index.html*

India Defence (2006). *The Indo-US Nuclear Deal: A Post-Henry Hyde Act Analysis. Http://www.indiadefence.com/analysis.doc*

India Semiconductor Association (2010). *Solar PV Industry 2010: Contemporary Scenario and Emerging Trends. Http://www.isaonline.org/*

Indiastat.com, various years and various states.

Institute for the Analysis of Global Security (2004). *India's Energy Security Challenge Energy Security.* January 21. *Http://www.iags.org/n0121043.htm ; http://www.southasia analysis.org/papers2/paper197.htm*

Intelli, Briefs (Blog) (2008). Full Text of the Nuclear Scientists' Letter to PM, IntelliBriefs: India. *Http://intellibriefs.blogspot.com/2008/06/india-ull-text-of-nuclear-scientists.html*

Iran–Pakistan–India Gas Pipeline, Wikipedia, the free encyclopedia. *Http://en.wikipedia.org / wiki/Iran-Pakistan-India_gas_pipeline*

Jeong, S.K. (2007). *Technology Trend Analysis of Capturing and Storing Carbon Dioxide*. Korea Institute of Energy Research (KIER). *Http://www.cheric.org/ippage/ip.php? code=p200704*

Kakotkar, Anil (2005). *Energy in India for the Coming Decades*, Lecture delivered at International Atomic Energy Agency,Paris. *Http://www.dae.gov.in/iaea/ak-paris0305.doc*

Kim, J.W. (2006). *China's Energy Consumption Status and Its Relating Policy*. International-Economic Research Institute of Korean Export-Import Bank. *Http://www.bricsinfo.org/ bricsinfo/trend/download2.jsp?seq=170*

Kim, J.Y. (2007). *China's Energy Policy and Strategic Implications*. The Korea Gas Union (KGU). *Http://www.kgu.or.kr/admin/data/U001 9a13a4e533eca4b805dba9a7833 fe205.pdf*

Kumar, Anand (nd). *Myanmar-Petrochina Agreement: A Setback to Indias Quest for Energy Security. Http://www.saag.org/ common/uploaded_files/paper1681.html*

Kumar, Pradeep and Amit Kumar Tyagi (2002). *Managing Energy Efficiently in Hotels and Commercial Buildings*. The Energy Research Institute.

Kumar, Sushil (2003). "Power Cycle Analysis of India, China, and Pakistan in Regional and Global Politics", *International Political Science Review* 24(1): 113/122. Sage Publishers, *http://www.jstor.org/stable/1601333Accessed: 17/06/2009 03:42*

Lal, Sumir (2005). "Can Good Economics Ever Be Good Politics-Case Study of Power Sector in India", *Economic and Political Weekly*: 649-656. February 12.

Lawrence, Ernest Orlando, Joanna Lewis and Ryan Wiser (2005). *Fostering a Renewable Energy Technology Industry: An International Comparison of Wind Industry Policy Support Mechanisms*. Environmental Energy Technologies Division, Berkeley National Laboratory, November. *http://eetd.lbl.gov/EA/EMP*

Leibenluft, Jacob (2005). "India's Nuclear Sell-Out", *Foreign Policy*, July, *http:// www.foreignpolicy.com/story/cms.php?story_id=3124*

Lee, M.S. (2008). *Wind Power Industrial Trend and Its Prospection*. Korea Development Bank (KDB) Research Institute, *http://www.kdbri.co.kr/*

Mahajan, V.S., S.K. Agnihotri and R.P. Athparia (1999). *Energy and Energy Resource Management*. Deep & Deep Publications Pvt. Ltd., Chapters. 2, 3 and 4.

Malone, Scott (2008). "GE Sees Big Nuclear Opportunity in India", Reuters News Agency. *Http://www.reuters.com/article/americasDealsNews/idUSTRE4917X720081002*

Marios Alexandrou, Project Management Fundamentals. *Http://www.mariosalexandrou.com /* project-management-fundamentals.asp?gclid =CIT14Kbpn6YCFcV66wod0wJ8MA

Ministry of Knowledge Economy (MKE) (2005). "New Renewable Energy" (White Paper). *Http:// www.mke.go.kr/ search/info/public/public_view.jsp? board_id=P_04_04_01&code =3700&sn=564*

Mukopadhyay, Kakali and Debesh Chakraborthy (2005). *Energy Intensity in India during Pre-Reform and Reform Period – An Input-Output Analysis*. Fifteenth International Input-Output Conference, the Renmin University (Beijing, China), June 27-July 1. *Http:// www.iioa.org/pdf/15th%20Conf/mukhopadhyay_chakraborty.pdf*

. Muni, S.D. and Girijesh Pant (2005). *India's Search for Energy Security Prospects for Cooperation with Extended Neighbourhood*, in association with Observer Research Foundation.

Navroz, Dubash K. and Sudhir Chella Rajan (2001a). *Politics of Power Sector Reforms in India. Collaborative Project on Power Sector Reform and Public Benefits in Developing and Transition Economies*. World Resources Institute and Inputs from the Tellus Institute. April 2.

———. (2001b). "Power Politics-Process of Power Sector Reform in India", *Economic and Political Weekly*: 3367-3391. September 1.

Pant, Harsh V. (2006). *PINR - The US-India Nuclear Deal: The End Game Begins*. The Power and Interest News Report. *Http://www.pinr.com*

Planning Commission (2006). *Integrated Energy Policy*, Report of Expert Committee, Government of India.

———. (2007). *Key Inputs for Accelerated Development of Indian Power Sector for 11th Plan & Beyond.*

———. (2002). *Power Sector Reform in Orissa: A Case Study in Restructuring.* Chapter 13.

Prayas (October 2001). *India Power Sector Reforms,* Update Issue 1, Pune.

Raj Energy (2006). *Indian Power Sector: Problems and Solutions. Http://www.rajenergy.com/a9slide.htm*

Ravindranath, N.H., N.V. Joshi, R. Sukumar and A. Saxena (2006). "Impact of Climate Change on Forests in India", *Current Science* 90(354): 16.

REN21 (2005). *Renewables 2005: Global Status Report*, REN21. *Http://www.ren21.net/pdf/RE2007_Global_Status_Report.pdf*

Sarkar, A.N. and Parag Diwan (ed.) (2008). "Energy Security", *Energycopia* 5, The Energy Research Institute, pp. 1-37, 44-54, 69-153.

Sengupta, Shuddhabrata (2008). *The Hyde Act and the 123 Treaty*. Kafila. *Http://kafila.org/2008/06/18/the-hyde-act-and-the-123-treaty-an-attempt-to-read-between-the-lines-of-a-noisy-debate/*

Shah, Chirag and Parag Diwan (ed.) (2008). "Energy Economics", *Energycopia* 1, The Energy Research Institute, pp. 1-10, 32-54.

Signitories (2006). "The Hyde Act and the Nuclear Scientist's Note", *The Hindu Daily. Http://www.hindu.com/2006/12/16/stories/2006121616171500.htm*

Smestad, Greg P. (2008). *The Basic Economics of Photovoltaics*. Solideas. *Http://www.solideas.com/papers/OSA_Solar_Stanford_Smestad.pdf*

Song, Joonki, Ryan Boas, Chris Bolman, Mark Farber, Hilary Flynn, Martin Meyers and Michael Rogol (2010). *The True Cost of Solar Power: Race to $1/W Photon Consulting LLC. Http://www.photonconsulting.com*

Squassoni, Sharon (2008). *Missed Opportunity: Nuclear Suppliers Group, Proliferation Analysis*. Carnegie Endowment for International Peace, September 2008. *Http://www.carnegieendowment.org/publications/index.cfm?fa=view&id=20448*

Taylor, Ian (2007). "Arms Sales to Africa: Beijing's Reputation at Risk Publication", China Brief 7(7), May 18. *Http://www.jamestown.org/programs/chinabrief/single/?tx_ttnews%5Btt_news% 5D=4069&tx_ttnews%5BbackPid%5D=197&no_cache=1*

The Business Standard (2010). "India's Hydrocarbon Profile Sees a Sea Change", *The Business Standard,* Tuesday, December 07, 2010. *Http://www.business-standard.com/india/news/india%5Cs-hydrocarbon-profile-seessea-change/363063/*

Thomas, Neff L. (1997). *Improving Energy Security in Pacific Asia: Diversification and Risk Reduction for Fossil and Nuclear Fuels.* Center for International Studies, Massachusetts Institute of Technology. Commissioned by the Pacific Asia Regional Energy Security (PARES) Project. December.

Tongia, Rahul (2003). *Power Sector Reform India: The Long Road Ahead.* Department of Engineering & Public Policy/School of Computer Science, Carnegie Mellon University.

Unnikrishnan, A.S. and D. Shankar (2007). "Area Sea Levels Trends along the North Indian Ocean Coasts: Consistent with Global Estimates?", *Global and Planetary Change* 57(301): 15.

Walters, David (2008). "US-India Nuclear Energy Agreement", *Dailykos. Http://www.dailykos.com/story/2008/9/9/192244/9707/224/592681*

Wee, J.H. (2008). *Reduction of Carbon Dioxide Omission Using CCS(Carbon Capture & Storage).* Environmental Management Corporation (EMC). *Http://www.emc.or.kr/include/board_union/union_filedown.asp?bbs_id=1879*

Weiss, W., I. Bergmann and G. Faninger (2004). *Solar Heating Worldwide: Markets and Contribution to Energy Supply 2001.* Paris: IEA Solar Heating and Cooling Programme. *Http://www.iea-shc.org/outputs/activities* iea_shc_solar_heating_ *worldwide_2001.pdf*

West, Larry (2005). *Sweden Aims to be World's First Oil-free Nation. Http://environment.about.com/od/renewable nergy/a/oilfreesweden.htm*

World Bank (2009). "Reducing Technical and NonTechnical Losses in the Power Sector", *Background Paper for the World Bank Group; Energy Sector Strategy.*

Others Sources

See Wikipedia articles on the list of nuclear reactors and India's nuclear reactors separately. The civilian nuclear reactors in India are located in Madras, Narora, Kakrapar, Kaiga, Rajasthan and Tarapur respectively.

Ministry of Petroleum and Natural Gas. *Http://www.petroleum.nic.in*

Director General of Hydro-Carbons. *Http://www.dghindia.org*